T0233033

# THE CHIEF DATA OFFICER MANAGEMENT HANDBOOK

## SET UP AND RUN AN ORGANIZATION'S DATA SUPPLY CHAIN

*Martin Treder*

**Apress®**

*The Chief Data Officer Management Handbook: Set Up and Run an Organization's Data Supply Chain*

Martin Treder
Königswinter, Nordrhein-Westfalen, Germany

ISBN-13 (pbk): 978-1-4842-6114-9               ISBN-13 (electronic): 978-1-4842-6115-6
https://doi.org/10.1007/978-1-4842-6115-6

## Copyright © 2020 by Martin Treder

This work is subject to copyright. All rights are reserved by the Publisher, whether the whole or part of the material is concerned, specifically the rights of translation, reprinting, reuse of illustrations, recitation, broadcasting, reproduction on microfilms or in any other physical way, and transmission or information storage and retrieval, electronic adaptation, computer software, or by similar or dissimilar methodology now known or hereafter developed.

Trademarked names, logos, and images may appear in this book. Rather than use a trademark symbol with every occurrence of a trademarked name, logo, or image we use the names, logos, and images only in an editorial fashion and to the benefit of the trademark owner, with no intention of infringement of the trademark.

The use in this publication of trade names, trademarks, service marks, and similar terms, even if they are not identified as such, is not to be taken as an expression of opinion as to whether or not they are subject to proprietary rights.

While the advice and information in this book are believed to be true and accurate at the date of publication, neither the authors nor the editors nor the publisher can accept any legal responsibility for any errors or omissions that may be made. The publisher makes no warranty, express or implied, with respect to the material contained herein.

Managing Director, Apress Media LLC: Welmoed Spahr
Acquisitions Editor: Shiva Ramachandran
Development Editor: Rita Fernando
Coordinating Editor: Rita Fernando

Cover designed by eStudioCalamar

Cartoons © by Timo Elliot

Distributed to the book trade worldwide by Springer Science+Business Media New York, 1 New York Plaza, New York, NY 100043. Phone 1-800-SPRINGER, fax (201) 348-4505, e-mail orders-ny@springer-sbm.com, or visit www.springeronline.com. Apress Media, LLC is a California LLC and the sole member (owner) is Springer Science + Business Media Finance Inc (SSBM Finance Inc). SSBM Finance Inc is a **Delaware** corporation.

For information on translations, please e-mail booktranslations@springernature.com; for reprint, paperback, or audio rights, please e-mail bookpermissions@springernature.com.

Apress titles may be purchased in bulk for academic, corporate, or promotional use. eBook versions and licenses are also available for most titles. For more information, reference our Print and eBook Bulk Sales web page at http://www.apress.com/bulk-sales.

Any source code or other supplementary material referenced by the author in this book is available to readers on GitHub via the book's product page, located at www.apress.com/9781484261149. For more detailed information, please visit http://www.apress.com/source-code.

Printed on acid-free paper

*For my family*

# Contents

# About the Author

**Martin Treder** is a seasoned Data Executive with 25 years of experience in international corporations. During the past decade, Martin established and led Data Management organizations at DHL Express, TNT Express, and FedEx Express International. He covered data areas as diverse as Data Governance, Masterdata Management, Data Modeling, Data Quality, Data Science, and Data Analytics.

While being a studied Mathematician (main topics: Operations Research and Applied Statistics), Martin has always focused on creating long-term commercial value through well-managed data and on shaping a data-driven culture.

At the same time, Martin is a sought-after speaker and panelist at international congresses, on topics around data, digitalization, and analytics. He also wrote the book *Becoming a data-driven Organisation* (Springer Vieweg, 2019), to convince leaders of the necessity to manage data actively.

# Acknowledgments

A universally applicable book cannot be written based on the experience of one individual. I would like to thank the global data community for sharing their challenges, approaches, successes, and failures through numerous conversations. The list of names would fill a book.

I should, however, mention the organizations that facilitate the necessary exchange among data professionals through conferences, events, and fairs.

I enjoyed the pleasure of being invited to speak at events organized by **Data Leaders** (UK), **IQPC** (United States), **ThinkLinkers** (Czech Republic), **Corinium Global Intelligence** (UK), **Vonlanthen Group** (Czech Republic), **GIA Global Group** (Czech Republic), **Encore Media Group** (UK), **Platinum Global Solutions** (UK), **Marcus Evans** (Czech Republic), **Cintona** (Switzerland), **Khaleej Times** (UAE), **Information Services Group** (United States), **Corporate Partners** (Czech Republic), **Hyperight AB** (Sweden), **Engaged Communications** (Germany), **DataCampus** (Germany), **The Economics Times** (India) and **IMH** (Cyprus).

It is always an enriching experience to contribute to your events and to learn from other speakers and attendees!

# Foreword

At a rather posh CDO conference, held in the shadow of the Duke of Northumberland's castle on the outskirts of London, I moved through the buffet line. As the Chairman of the event, I dutifully chatted with anyone and everyone, but at this very moment, I was a bit hungry and a little jet lagged. While keeping uncharacteristically to myself, scooping up a particularly delectable piece of chicken, I was startled by a booming voice from the opposite side of the serving dishes.

"I'll bet you didn't like what that last speaker said?"

In my role, I would need to be diplomatic and softly uncritical of any presenter. "Pardon me?" I answered.

"I say, I'll bet you didn't like what that last speaker said: that most CDOs should eventually evolve into CD-A-Os!"

That comment had hit a nerve during the session, and this fellow chicken scooper knew enough about me to know which nerve it hit.

"No!" I barked, "I did *not!*"

"I was sure you did *not*," he smiled.

And so began my immediate friendship with Martin Treder. We realized, him before me, how like-minded we are, especially when it comes to the importance of data.

I am Scott Taylor, known as The Data Whisperer. I help calm data down. That is what we all do in the data management space: calm data down. Data is unruly, data is big, data is unstructured, it needs to be calmed down.

Like Martin, I am a firm believer in the strategic importance of proper data management. We share a mutual understanding of what is necessary to manage data. We understand, to our core, how data, especially foundational data, can help an enterprise grow, improve, and protect its business. I also believe that things like analytics, as important as they are, get a disproportionate amount of attention and limelight. It is the cool, fancy, sexy stuff. So that makes me an advocate for the unsung, relatively dull, but critically important stuff. There is no analytical insight without a proper data foundation.

We have both been on this mission for a while, evangelizing to organizations that they need to pay attention to their data – not just the colorful charts and reports, but getting the basics right. From my perspective, data and data management are about determining the truth. Analytics and business intelligence are about deriving meaning. I would suggest that most data experts already know this. They support it. They understand it. They believe it. In fact, no data expert I have ever met thinks data does not beget more value when it is properly used in analytics. Martin and I understand that too. Thinking about data and analytics, they work together. They do need each other. These two things are linked, but it helps to think of them separately. "At every Enterprise there is," as Martin puts it, "the need for a solid data foundation for all the fancy stuff that people intend to do with data."

Today's macrotrends are about the "fancy stuff": All about Cloud, about data science, about analytics on the edge, and other cool things. Those all still need a solid data foundation. No matter where we have been, no matter where we are now, no matter where we are going, whatever the future is, you are always going to need "a solid data foundation for all the fancy stuff."

I rally for the "unfancy" stuff and sometimes unsung *data* professionals in data management, data governance, MDM, RDM, masterdata, metadata, and the like, who know every day that their work is foundational, is bedrock, and enables the ability for an enterprise to flourish with gloriously informed analytics.

If you want a healthy organization, the lifeblood that goes through your organization is your data, and you have to work on it. It is like trying to lose weight and get in better shape. You have to eat differently. You have to exercise. You have to understand and master the basics. You have to do the hard work. There is no magic out there.

I hope you can tell by now that I am good at "talking about data." I focus on the "strategic WHY" rather than the "practical HOW." I assume since you are reading this book, you will need more than simply a good story to share in the elevator when you meet with your CEO. You will need practical advice from a seasoned data professional – someone who has put in the time, made the mistakes, overcome the challenges, celebrated the successes, and understands the current and future role of CDOs. That is why *The Chief Data Officer Management Handbook* will become your valuable go-to resource. Martin delves into his vast expertise, and the experience shared by other professions, to help you understand the design, execution, and even the psychology of the most important addition to the C-suite in decades – THE CDO. (And note, no "A"s added in there.)

What Martin delivers is advice you can put into action, not academic theory. This is a *handbook*, not a textbook. It is simple, but not simplistic. It will help you develop the skills you need beyond your technical knowledge. Let this be a helpful guide during your daily business life as a current (and even aspiring) Chief Data Officer.

I have seen Martin present to a room of data leaders. He can whip up a crowd with his unbounded enthusiasm. He can inspire them with his sheer determination about the value and power of data. At times he seems ready to burst through the podium, roll up his sleeves, and dig into your data himself. My hope is you can experience that energy firsthand someday; until then you have in your hands an excellent substitute. Read this book and you will get more than enough. And maybe, someday, you will be lucky enough to see him inspire in person or even share a simple chat across a buffet line. Enjoy!

Scott Taylor
*The Data Whisperer*
*MetaMeta Consulting*

# Preface

You have opened this book for a reason.

You might have been asked by your organization to bring its data handling in order. Your CEO may have become furious upon finding out that certain data is not available or that forecasts were precise but wrong.

Or, watching how your organization deals with data, you might have thought that there must be a better way.

Either way, you are touching on one of the most underestimated business topics of our times:

**The need for a solid data foundation for all the fancy stuff that people intend to do with data.**

Have you ever wondered why so many organizations call data the "oil of the twenty-first century" and at the same time struggle in becoming truly data driven?

There is no simple answer to this question. Different issues contribute to this problem to varying degrees.

But understanding those factors is critical if you want to help your organization use data in a commercially successful way, both short and long term.

I invite you to dive into this book and read what some of the pioneers of this discipline have found out.

But, more importantly, please reflect on what you read in the light of the organization you are working for.

Hundreds of pages about Data Management are too much to read, considering your limited spare time? Don't worry, this book is not a novel! You can start wherever you want, pick the topics you are most interested in and decide on your individual reading order. You will find a few cross-references, but none of the chapters requires you to read previous parts of the book.

Enjoy!

# Introduction

The Chief Data Officer (CDO) is on the rise.

Already in 2018, Randy Bean[1] wrote on Forbes.com:

> While only 12% of executives reported that their firm had appointed a Chief Data Officer when the survey was first conducted in 2012, there has been a sharp and steady increase in adoption of this new c-executive role over the course of the past several years. In the 2018 survey, which was recently released, nearly two-thirds of executives – 63.4% – now report their firm having a CDO. Clearly, the Chief Data Officer has become an established role within a majority of leading corporations.

And in March 2020, Gary Richardson[2] reported a further increase:

> The role of the Chief Data Officer (CDO) is evolving and gaining traction within the industry – 70 per cent of companies have now appointed one, up from 12 per cent in 2012. The role's success is crucial for business growth, and it is clear that CDOs must develop their role away from a defensive data strategy and embrace their roles as agents of change.[3]

At the same time, however, data-minded people still seem to face difficulties trying to effectively develop their organizations toward a data-driven mindset.

I made it a habit to listen to the concerns of data experts that work for all kinds of organizations.[4] More and more of those originally engaged change agents are reporting a rising degree of frustration. Most of them came to the conclusion that it is almost impossible to introduce data-centric thinking in organizations from the bottom up.

---

[1]Randy Bean is the CEO and managing partner of consultancy NewVantage Partners.

[2]Gary Richardson is the MD for Emerging Technology at 6point6, leading a team in the development of AI and Machine Learning solutions.

[3]Gary Richardson, "CDO – you got the job, now how do you avoid getting fired?" ITProPortal, March 2, 2020, www.itproportal.com/features/cdo-you-got-the-job-now-how-do-you-avoid-getting-fired/

[4]Great opportunities to listen and learn are data conferences and experts groups on professional networks.

But even at the top, where you would expect sufficient authority to drive change, the situation seems to be challenging. As recently as February 2020, Professor Michael Wade[5] stated that the average tenure of a CDO is 31 months, shorter than other C-suite roles.[6]

# Why is that?

In most cases, the issue is not around their concept or about their knowledge. You can assume most of these CDOs to be seasoned data experts.

Understanding the value of data and the science behind it is indeed helpful. But it is not sufficient. The challenge is to get people engaged – both the executives and the folks on the ground.

A CDO without formal authority is a toothless tiger. But power alone will not make a CDO succeed.

The skills a CDO needs to develop can usually be divided into the following three categories:

1. Data expertise

2. Organization skills and business orientation

3. Diplomatic and communicative skills

Hardly any CDO is weak at the first point. CDOs usually keep abreast of technological development and know what they are talking about.

While a Data Management book would be incomplete without covering parts of this category, the main focus of this book is, therefore, on the other two points.

For a CDO to succeed, I suggest a three-step approach:

1. Understand your organization.

2. Develop your target structure.

3. Set up and follow your implementation plan.

---

[5]Michael Wade is Professor of Innovation and Strategy, Cisco Chair in Digital Business Transformation at IMD Business School, Lausanne/Switzerland.

[6]Michael Wade, "From dazzling to departed - why Chief Digital Officers are doomed to fail," World Economic Forum, February 12, 2020, https://www.weforum.org/agenda/2020/02/chief-digital-officer-cdo-skills-tenure-fail/.

This book intends to support you on this journey. It is divided into three main parts:

- The first part (Chapters 1–11) is about *Designing an Effective Data Office*. It covers the structure of a data management organization, in response to today's gaps and opportunities.

- The second part (Chapters 12–15), *The Psychology of Data Management*, addresses typical challenges in converting a company into a data-driven organization.

- The third part (Chapters 16–23) is captioned *Practical Aspects of Data Management*. It deals with typical topics of a Data Office.

Throughout the book, I use the acronym CDO as an abbreviation for Chief Data Officer. Your role may be called differently, as no de facto standard has been established yet. But even if you are a *Business Information Manager*, a *Chief Data and Analytics Officer*, or a *Head of Data Management*, please keep reading. The title should be the least of our concerns.

# Designing an Effective Data Office

Companies that have not yet built a data strategy and a strong data management function need to catch up very fast or start planning for their exit.

—Thomas H. Davenport, "What's your data strategy?"
(Davenport, 2017)

# Understand Your Organization

*"Watch out! Here comes the intelligent enterprise!"*

**Figure 1-1.** The intelligent enterprise

© Martin Treder 2020
M. Treder, *The Chief Data Officer Management Handbook*,
https://doi.org/10.1007/978-1-4842-6115-6_1

Everybody talks about "data." When you ask a company executive about the importance of data, the typical response is violent agreement. In fact, an increasing number of organizations realize that data is gaining more relevance every day.

Yes, they have been successful for years if not decades without dedicated data management. But it doesn't require a data expert to see how the evolution of tech has made data management imperative. There are many opportunities that come with data, and there are many competitive disadvantages to those who do not engage with it.

Does this mean that most organizations are well prepared to tackle both the challenges and opportunities that come with the age of data?

You will have guessed that this was meant to be a rhetorical question. But what prevents organizations from taking that critical step from awareness to action?

Let's have a closer look.

To change an organization, you need to understand the organization.

To understand how organizations deal with data today, it pays to look at the historical development: Data did not enter the scene all of a sudden. The relevance of data has developed gradually over the past few decades.

As a consequence, there was often no single moment or event that prompted an organization to decide "we need to govern and coordinate data." Just as the boiling frog in the fable,[1] organizations that would have reacted to a big bang appearance of data failed to realize the gradual ascent of data.

Instead, more and more good people in various areas of an organization saw the need or the opportunity to act. Often, the IT department became the incubator, where people realized the technical opportunities. In many other cases, though, nontechnical people in business[2] departments felt the pain and decided to do something about it.

Over time, this development has led to a number of typical approaches (and resulting issues). Let us have a closer look.

---

[1] A good description can be found under Wikipedia – Boiling Frog, 2019.

[2] I am using the term "business" across this book to describe entities benefiting from well-managed data. This is, of course, not really accurate as it suggests a clear distinction between data service providers and data service consumers. In fact, while IT is mostly a service provider in this context, it is also a consumer where data helps manage the IT "business."

# Five implicit Data Governance models

You cannot NOT govern data. Data is all around you, and people will deal with it. They will do whatever seems best in their particular situation, within the framework of what is allowed and to what extent their organization sanctions non-allowed activities.

This situation is similar to that of political states and their system of government, which allows me to describe five models of organizations' approach to data using political labels (Figure 1-2).

We will see that what works best as a form of state government does not necessarily work best in organizations. This finding applies particularly to broad topics such as data management.

Feel free to search for these models within your organization. Spotting pattern helps you address the related risks.

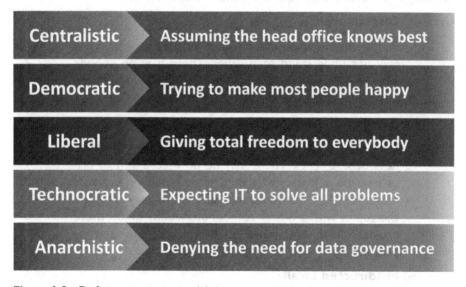

| Centralistic | Assuming the head office knows best |
| Democratic | Trying to make most people happy |
| Liberal | Giving total freedom to everybody |
| Technocratic | Expecting IT to solve all problems |
| Anarchistic | Denying the need for data governance |

**Figure 1-2.** De facto governance models in organizations

# Centralistic Data Governance

Centralistic models tend to develop in organizations with a tightly managed structure. You often find them in traditional brick-and-mortar organizations. I sometimes also call this the "ivory tower model," as it often fails to consider local knowledge and experience.

In many cases, the positive starting point has been an organization's conviction that, in all situations, "somebody must be responsible."

In real life, this approach often leads to the following situation, when it comes to the handling of data:

- All data experts of an organization belong to one central team.

- Other departments are expected to go to this team for insight. Departmental activities are forbidden or don't get funding.

- The team's focus is on Analytics only. Some centralistic organizations also have a central team that deals with Masterdata, but the two teams are usually not forced to work together.

This very traditional setup, however, comes with a few risks:

(i) **Disconnect**

People are either business experts or data experts. They may not understand each other. There may be nobody who understands both the business problem and the options to address it through data management.

(ii) **Conflict of interest**

Data folks often have different objectives than the business departments. There may be no mechanism to prioritize the activities of a data team based on business relevance. As a consequence, data teams may focus on the topics with a higher success probability or on those projects that are more fun from a data science perspective.

(iii) **Misdirected loyalty**

Where you have a clear distinction between the business teams and their central "service providers," people don't feel part of a bigger group. They easily think in terms of "them" (the other side) and "us." As a consequence, data folks often do not feel part of the business decisions taken based on their input, and they don't see the link to their work. And whenever business teams are successful, they will tend to celebrate with everybody they consider part of "us" – which often excludes those "service providers."

All in all, the centralistic model is no longer suited to support a business world that increasingly depends on cross-functional collaboration.

# Democratic Data Governance

Democratic Data Governance sounds good, right?

We should be cognizant of the fact that democracy was developed for communities such as political states. Most members of such communities are long-term stakeholders and thus less interested in short-term gains.

To make the step from the raw model of participation to sustainable democracy, those communities had to introduce certain rules to avoid short-term gains for majorities that would negatively impact the entire population in the long run.

---

### EXAMPLE

Let's imagine a community on an isolated island. In an attempt to improve their communal life, they agree to take all decisions in a democratic way.

After a while, the poorest three-quarters of this community get organized. They propose to evenly distribute the wealth of the wealthiest quarter among the entire population.

Unfortunately, the island community does not have a "rule" that protects property. As a result, a law to redistribute wealth is passed as 75 percent of the population voted in favor of this proposal.

From that moment on, however, people tried to avoid becoming part of the wealthiest quarter of the population. Productivity and creativity stalled, and the island's prosperity was hit severely.

---

Most democracies managed to agree on a set of rules to avoid such situations.[3] They understood that what is immediately beneficial to a majority may jeopardize the community as a whole in the long run.

While such a setup is excellent for states, it has turned out to be suboptimal for organizations. Here, managers may target at selfishly increasing the success of their own department by gathering the agreement of a majority. If their proposal results in small benefits for two-thirds of the organization while being a disaster for the remaining third, they would still be able to obtain a majority agreement, without doing any good to the overall organization.

---

[3] In modern forms of government, such a set of rules is called the "constitution."

At best, such managers can hope for the organization to be big enough to survive their selfishness. At worst, they can leave the organization at any time, with a seemingly impressive success added to their record.

So, what is a typical setup of a democratic organization?

- All departments are part of the dialogue (which is definitely a good thing!).

- Agreement (sometimes even formal approval) is always sought from all stakeholders.

- The main criterion for a decision is a consensus or a majority vote.

This setup comes with quite a few risks:

- "Agreed by a majority" is not equivalent to "the best solution." A majority vote may represent a minority of value, resulting in an overall negative impact on the entire organization.

- Personal agendas can block an organization's way forward.

- Managers are rewarded for agreeing to colleagues' proposals so that these colleagues agree to their own suggestions later.

- A democratic process may come with long lead times before a decision can finally be taken.

- Missing support of the majority may be used as an excuse for inactivity: "I have tried everything, but the organization didn't want to follow."

It is a significant danger of this model that it sounds like the right thing to do. Who would dare say something against democracy?

## Liberal Data Governance

Some organizations decide not to regulate the handling of data.

Interestingly, we are not primarily talking about traditional organizations that are strangers to modern data topics such as digitalization.

Instead, we see a considerable number of high-tech startups apply this model. The reason in most cases is that these organizations focus on developing their main business area at high speed, and they don't want any internal regulation to slow them down.

This behavior reminds me of the protagonists of the first dot-com wave. Many of them failed due to the absence of a proper financial setup. The lack of a CFO back then is equivalent to the lack of a CDO today. And please expect the consequences to be similar!

Liberal data governance typically comes with the following characteristics:

- Total autonomy for all parties.

- Responsibilities in data matters defined at departmental level, to varying degrees.

- Many pseudo-experts in data matters.

- Wherever people publicly exchange data approaches, their focus is not on convincing others but self-praise.

Such an environment comes with an extensive list of risks:

- Lots of insular setups (e.g., multiple independent, mutually incompatible glossaries).

- Slow business model changes.

- People see issues but don't know whom to talk to.

- Half of the people become fatalists. The other half become egoists (slightly exaggerated).

- The complexity of change: Nobody has an overview of the entire impact of a particular change. That is why significant business changes tend to require a full assessment of the IT landscape.

# Technocratic Data Governance

At a data conference in Sept 2019, I heard someone say: "The solution to our data problems is not technology, it's architecture!" Needless to say that this person worked for his organization's Architecture team…

While I agree with the first part of the statement, the second part reflects a typical perspective: The assumption that you need to start with the right part of the *solution*. A technocratic approach often forgets that, instead, you should start with the *problem*.

My advice for technocratic organizations: For data to be governed effectively, *all* aspects need to contribute – but please consider the sequence, and don't start with technology. (Note: Don't start with architecture, either!)

But what does a technocratic organization look like? We typically find this setup:

- People perceive data as a technical topic.

- Responsibilities in data matters are clearly defined – but it's all with IT.

- Lots of software is in place to manage and analyze data – but hardly anybody on the business side understands the background.

- Approach: "Here's the solution. Do you have a problem to solve?"

Risks:

- Opportunities are missed for which *both* the business situation and the technical options need to be understood.

- There is an increasing disengagement on the business side: "Let IT deal with it."

IT often don't even perceive the situation as an issue. "It works…"

## Anarchistic Data Governance

Often, business veterans get used to trusting in their decades of experience. Capitalizing on the experience and knowledge of many such veterans is typical for traditional organizations.

The usage of experience and gut feeling is often well governed in such organizations. You find plenty of steering boards – where managers exchange their views and somehow come to a decision. Yet nobody has ever tried to describe the road "from guts to mind": How to validate feelings, how to put them into perspective, how to match them with tangible business objectives?

Instead, whatever an experienced manager says is assumed to be the result of a sound (yet subconscious) evaluation process executed by that manager.

As you will guess, there is no place for data in such a model, other than on an anecdotal basis. (Would you count "I have already seen it happen twice this month!" as using data?)

Hardly any conscious development of governance would result in such a data governance model.

And, indeed, anarchistic data governance does not get introduced as a result of a decision. It rather stands for the absence of any consciously introduced model.

Unfortunately, this is precisely the situation many organizations are in when it comes to the management of data!

Anarchistic data governance typically comes with the following characteristics (and I am only slightly exaggerating):

- Data is a nontopic outside regulated areas.
- Respect for managers is proportional to their seniority. Someone with 30 years of service must know.
- "John said so!" is considered a valid argument.
- It is difficult to comprehend management decisions.
- Implementation of changes to the business model is risky as nobody dares question executive decisions.

This governance model has one essential risk in common with the liberal model:

- The complexity of change: Nobody can estimate the impact of change, as there is no overview of the data landscape. That is why any bigger business change requires a full assessment of the IT landscape.

It is difficult in such an environment to learn from failures through root cause analyses. Instead, people tend to avoid decisions that have led to failure before, without asking for the causality.

www.timoelliott.com

*"I'm making a decision! Stop confusing me with facts!"*

**Figure 1-3.** Facts as the enemy of decisions

## Behavioral patterns in data matters

It is important to understand the current data governance approach in your organization. Concrete action, though, requires a deeper understanding of how (and why) people in your organization react to the data challenge.

### "Data is an IT task"

If you ask 100 arbitrarily chosen employees of any organization "Is data an IT topic?", you would hear "yes" at least 95 times.

Why do people think so? After all, data had existed long before there were computers.

You would like to hear an example? Here you are!

The bible starts the description of the birth of Jesus Christ with the following words:[4]

> 1 In those days a decree went out from Caesar Augustus that all the world should be registered. 2 This was the first registration when Quirinius was governor of Syria. 3 And all went to be registered, each to his own town. (Luke, 1971)

---

[4]Luke 2, 1–3, Holy Bible, English Standard Version

This census is an ancient example of data management. We can assume that Caesar Augustus did not have an IT department. He might well have had a **data** department, though.

Let's have a look at the role of IT. Since this discipline developed, it has helped handle all of that data more effectively, using hardware and software.

As a result, whoever wanted to get any data had to talk to IT. The same applied to the capturing of data and the transmission and storage of data.

And IT stands for tremendous progress! We could only use a fraction of the value of our data without today's technology. Caesar Augustus would have been excited to have at least some of those possibilities in processing the data he gathered.

What many people seem to have forgotten: IT stands for Information *Technology*, not for the information itself. IT is supposed to provide services to the business functions which own that data, to ease the handling of data.

Can we now conclude that we need to take "data" away from IT?

In fact, it is not vital to the success of a Data Office, whether it sits under IT or not. It is the **mandate** and the support that counts. And a Data Office needs to be a distinct team, not mixed with any other discipline.

Under these circumstances, the Data Office should reside where its credibility and effectiveness are highest. This may be the IT department or any other function, depending on the culture and tradition of the organization.

Notwithstanding the preceding statement, you may wish to let the workforce know that data is *not* Information Technology.

# "We can focus on Analytics"

The data world is too big and too interconnected to concentrate on one single aspect. You need to look at the entire data universe to achieve sustainable management of data. But you will find again and again that people use "data" and "analytics" as synonyms.

The situation reminds me of this old story:

*One day the hen gathers her chicks and tells them solemnly: "Now that you are old enough, I can share a piece of wisdom with you. So far, you got told that the fence of the chicken coop is the end of the world." She pointed to the horizon with her wing, took a deep breath and shared her wisdom with the next generation: "The fence is **not** the end of the world! The world only ends over there, at the edge of the forest…"*

If an organization's data activities start directly with Data Science and Analytics, you might not obtain the desired results. This is because the data world reaches beyond Analytics.

This is why you may wish to create a proper **data foundation** first, to ensure the structure, verbiage, and quality of the data used are well understood. Such a foundation needs to cover both Masterdata and transactional data, and it should apply to both internal and external data. Properly defined data governance is required to ensure everybody plays by the same rules.

If you compare data with an iceberg, Analytics is the visible part. Remember, this is by far the smaller part!

But how can you tell whether an organization has a comprehensive view of data, or whether it focuses on Analytics?

Have a look at the role description of the organization's top data lead. Even if the role is called "Head of Data Management" or "Chief Data Officer" – if the description is that of a Chief Analytics Officer or of a Machine Learning team lead – the organization may get it wrong.

A typical alarm signal is an organization's expectation that a Chief Data Officer must be able to guide the data scientists technically, for example, in implementing unsupervised Machine Learning models using Python and TensorFlow. Such an organization is indeed confusing a CDO with a Data Science team lead.[5]

This is a dangerous situation if not resolved early, as the gap does not become visible too soon because such a data team will keep producing "results."

## "It's digitalization"

Digitalization is one of many areas that require proper data management. But it is not the only one. As a consequence, the priorities in managing your organization's data should not be set by the organization's digitalization team (nor by any other single area that requires data).

Digitalization may turn out to be your Data Office's number one topic. But you can only tell after a cross-functional review of all areas!

---

[5] A mature organization will require both roles. And even if the Head of Data Science reports into the CDO, the latter will not necessarily have a higher salary, following the rule of demand and supply.

# Paralysis by analysis

Some organizations are convinced they are data driven while, in fact, they are data hindered.

I am talking about organizations which allow the misuse of data. As a result, data becomes a weapon, or people use data to question each decision they dislike.

As you will know, you can prove almost everything wrong using data – if it is you who decides which data to use and how to do it.

But most frequently, people don't dare decide without having 100 percent data support. A typical case is a choice between "Go for X" and "Do not go for X" where data can prove that both options come with risks. Data-hindered organizations tend NOT to decide in such a case. Instead, the requestor will be asked to reassess and to come back with more evidence. Eventually, no decision will be taken at all, or the requestor will tweak the data so that it fully supports one of the options. Obviously, neither outcome is desirable.

Such behavior provides a good case for a Data Office. Given its neutrality, it has no interest in pleasing any of the parties.

This is also a good reason for organizations *not* to have more than one (top-level) CDO: Someone must have the final word.

# "Digital Natives know how to do it"

Do they, really?

Most Digital Natives are predominantly familiar with the *usage* of data. They do not necessarily know how to acquire, manage, organize, and provide data.

Here we are facing two fundamentally different skill sets. You don't become a starred chef if you know how to eat at the Queen's table.

The biggest issue among the Digital Natives is often their simplified view of data. As they are users of modern devices, they tend to see only the easy parts of it. Or as *c't*, Europe's leading magazine for IT and technology, put it, "While all Digital Natives know how to post a selfie on Instagram, hardly any of them knows what a command line is, let alone being able to program" (Gieselmann, 2020).[6]

Instead, we need people who know how to serve those Digital Natives. Unfortunately, this is a rare species.

---

[6]Original German text: "Die Digital Natives können zwar alle Selfies auf Instagram posten, aber kaum einer weiß, was eine Kommandzeile ist – geschweige denn, wie man programmiert."

## "Our business functions can do data on their own"

Functional experts are often assumed to be able to deal with the data aspects of their work as well. This assumption is not only wrong in many cases. It also leads to **silo thinking**, inconsistencies, and double work.

## "It is all good"

This is by far the most frequent statement I have heard executives say. They ask "Which problem do you intend to solve?"

You hear this statement frequently in traditional organizations that have been successful for decades despite not even thinking of dedicated data management.

But you also find it in dynamic startups who feel well prepared to do effective reporting, data analysis, and AI.

This case is the most challenging one. The question itself is justified. But it will remain unanswered if the person asking it is not willing to listen – to you or to any other person considered a trusted advisor.

## "Tidy up and tick the box"

Not all organizations fail to realize that they face a data issue. The impact of bad data has become too obvious, or required data is simply not available.

But not all responses are equally well suited to solve the issues sustainably. A very dangerous choice is the "one-off" approach: Setting up a project to address an organization's data problems and intending to return to business as usual after project closure.

What does such an approach look like? Such an organization would not search for a Chief Data Officer or even set up a Data Office. Instead, they will engage an external consultancy firm or search for a temporary data manager to "solve our data issues."

As you may guess, this way of thinking is doomed to fail, just as you cannot repair a technical device "for good" and not expect it to break again.

It is, therefore, crucial to embed the knowledge in the organization that data must be taken care of similarly to other assets – which implies the need for permanent active management.

Dr Jürgen Schubert, Director Master Data Governance at Infineon Technologies, put it this way: "Data is not a project with an end-date. You need to invest energy permanently."

But why is it that the same organizations that would never assume they can stop maintaining their tangible assets think they can tidy up their data once and for all?

During the past few years, I have come across three main patterns:

(i) **Data is not seen as an asset.**

The first reason is undoubtedly that data is not yet seen as a full-value asset.

In such cases, I recommend the explicit comparison with the organization's traditional (tangible) assets, as well as a closer look at other intangible assets such as patents. People will find more commonalities than one would think of at first glance. As a result, such a comparison usually has a surprise effect.

See Chapter 16 for additional thoughts on data as an asset.

(ii) **Tactical data handling.**

Some organizations follow the pragmatic approach of "Whenever it is broken, we will fix it!" They are willing to launch another data fixing project as soon as it becomes necessary again.

If you face this situation, it may help to quote external expert knowledge. It was as early as 1993 (hardly any-body has talked about data management back then) that G. Labovitz and Y. Chang described the **1-10-100 rule** in their famous book *Making Quality Work: A Leadership Guide for the Results-Driven Manager* (Labovitz, Chang, & Rosansky, 1993). This rule quantifies the hidden costs of poor quality.

According to the 1-10-100 rule, you can reduce the cost of failure by factor ten if you correct the error before it causes damage. And you can even increase the cost reduction by factor 100 if you prevent a mistake from happening in the first place.[7]

---

[7] You will probably have come across the 1-10-100 pyramid that is often used to illustrate this rule.

Even if there is no scientific evidence for the precise factors, the order of magnitude of this rule has been confirmed over time in various areas, including that of Data Quality.

In other words: Yes, ongoing data management costs money. But if it is used to keep your Data Quality in good shape, your organization will save a lot of money compared to the need to fix the quality of data, and you will save even more money compared to the damage caused by insufficient Data Quality.

It should not be too challenging to underpin this rule with examples observed in your organization.

You read more about Data Quality in Chapter 10.

(iii) **Data is considered a competitor.**

Another, entirely different reason for nonsustainable approaches is current executives' fear to lose influence or independence. They'd rather engage an external team that can be sent home at any time than accepting a permanent additional authority.

The challenge with this situation is that nobody will admit being driven by such motives. Whenever you have the feeling that an executive "should know better," you may wish to dig deeper and ask further questions to understand the real drivers of hostile behavior.

If you position the Data Office as a permanent source of support, you may be able to illustrate its immediate value. People need to understand that a Data Office's main target is to help other functions perform better, not to replace them.

You find further thoughts around stakeholder management in Chapter 4.

# Aspects of Effective Data Management

*"When you two have finished arguing your opinions, I actually have data!"*

www.timoelliott.com

**Figure 2-1.** Opinions are good – data is better

© Martin Treder 2020
M. Treder, *The Chief Data Officer Management Handbook*,
https://doi.org/10.1007/978-1-4842-6115-6_2

# Maturity assessment

Many different aspects need to be considered for data management to become effective. This chapter intends to give you an overview of the most critical elements before subsequent chapters will allow for a deep dive.

You may wish to perform a **maturity assessment** of your organization vs. the "ideal state" as you read through this chapter. I recommend that you do not spend too much effort on formalizing this assessment. It is sufficient to be able to measure so that you can quantify progress as you proceed.

It is essential, however, to understand that there are two different target groups for a maturity assessment, requiring totally different approaches:

- The first target group consists of yourself and everybody who is already fully committed to having your organization become data driven. This target group needs to know how much remains to be done where and how high the resistance is going to be.

- The second group consists of all stakeholders who still need to be convinced that "we have a problem today." For this group, a gap analysis would contain missed opportunities and unaddressed issues, together with the respective financial impact.

You would ideally work on both assessments in parallel. You would not use the first one to your stakeholders before they are convinced. But as soon as they are, this maturity assessment is a sound basis for a roadmap and an open discussion on priorities.

# The two main gaps

As we have seen in the previous chapter, none of the frequently observed Data Governance models seems to adequately address the challenges and opportunities that come with data management and analytics.

So, what are the main gaps across all models?

Most organizations face the following two challenges:

- **Lack of collaboration**
- **Lack of business ownership**

I recommend that you have a close look at these two areas as you assess your organization. They will pop up over and over again, and they should be addressed across all data management disciplines.

# Subsidiarity

We have seen that historically developed ways of dealing with data can be described in political terms. But as we are trying to develop useful **solution** approaches, we find good analogies in the area of state government as well.

One recipe that has proven successful in both worlds is *subsidiarity*.

**Subsidiarity** is "the principle that a central authority should have a subsidiary function, performing only those tasks which cannot be performed at a more local level."[1]

But what does this principle mean in the context of data management?

You often find leaders ask themselves:

- Should data be managed centrally by an empowered, highly specialized central team?
- Or should data management be left to everybody across the organization?

If you follow the subsidiarity principle, you will find that it's **none** of these two extremes.

Instead, subsidiarity suggests that you centralize data responsibilities where it makes sense and that you leave the rest to the "field" – which may stand for business functions, subsidiary firms, or geographical units, depending on your organization.

Splitting responsibilities intelligently alone will not solve your problems. To make it work, you need to equip all involved teams with everything they need!

Throughout this book, you will find recommendations as to which areas of Data Management should be delegated or centralized. For now, we can state our first theorem.

---

[1]Oxford English Dictionary; https://en.oxforddictionaries.com/definition/subsidiarity

```
┌─────────────────────────────────────────────────────────────┐
│              DATA MANAGEMENT THEOREM #1                      │
└─────────────────────────────────────────────────────────────┘
```

The organization of data-related responsibilities and activities demands a careful balancing between centralization and delegation.

- Any centralization requires good reasons.

- Any delegation must involve trust and support.

# Business orientation

Technology-driven people are often tempted to start with a solution. The idea is that "there must be an application for this fancy new piece of technology!"

If you want to avoid the trap of unused high technology, don't start with the solution – start with the issue or opportunity instead.

Experience has shown that this paradigm shift is hard to digest, particularly for seasoned IT leaders. They are used to offering state-of-the-art solutions proactively.

To address this challenge, you may consider the following three recommendations:

- Always ask: Is there a problem (something doesn't work as it should), an opportunity (it works, but improvement is possible) or an innovation (a new business model)? Determine and quantify the real pain or possible gain.

- Seek the dialogue: Jointly hypothesize about benefits, and validate them together with the respective business functions.

- Develop a product mindset: The shift from project manager to product owner is beneficial for data management. It adds the long-term view, beyond the end of a project!

I am using two diagrams to illustrate the change. The suggested increase in business input is reflected by an increase in light color in the second diagram.

This is how IT teams have worked for decades.

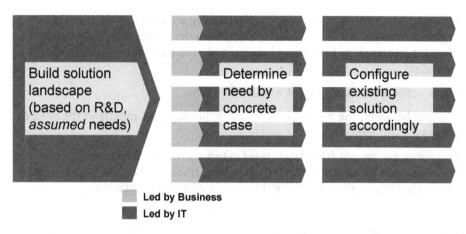

**Figure 2-2.** The traditional sequence of steps

And this is the new model of involving business functions early.

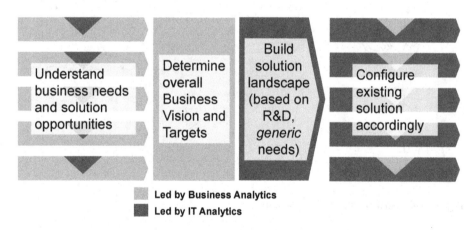

**Figure 2-3.** The business-driven approach

An important aspect of this model is the lack of a strict handover point: Both sides are part of the dialogue throughout the entire cycle. This is even possible for a waterfall project or throughout the whole lifetime of a product.

Very small companies usually don't face a lack of business orientation. Here, IT people are often close to the business, for simple reasons of size. Sometimes the same person that optimizes a process is also responsible for the subsequent modification of the software.

However, this admittedly very effective approach is not scalable. As organizations grow, the need for specialization increases. But a specialization of roles requires a consciously organized dialogue between all of these specialists.

Here, active data management enters the scene.

The gap between the business view and technology lies mostly in the area of processes and data.

Furthermore, people from both sides often consider themselves open while unconsciously being stuck in their old way of thinking. And many of them are convinced they think and act for the benefit of the organization.

**Figure 2-4.** Analytics too busy for analytics

To address this situation, a Data Office requires team members who

- Understand processes and data

- Understand what drives both IT specialists and business professionals

- Manage to remain "neutral," that is, they don't work with a predefined perspective and are equally open to both sides' ideas and concern

- Can moderate and show empathy

Again, the mandate for such a Data Office role must come from the top, while respect and credibility need to be achieved through good work of the Data Office.

---

## DATA MANAGEMENT THEOREM #2

Data Management is everybody's job – throughout the entire organization. It is particularly not an IT job, and it does not start with technology.

It is about building bridges between business and IT, based on a solid understanding of both sides.

---

# Commercial orientation

Unless you work for a public or non-for-profit organization, dealing with data needs to add measurable shareholder value.

Eventually, any organization that invests in data wants to see a return. I call it *Return on Data*.

This expression is derived from the financial term "Return on Investment" (RoI) – which is critical for any activity that costs money: You do it only if it pays off.

The idea behind the expression "Return on Data" is that dealing with data is not an end in itself. All data activities must be judged by the effect they have on the organization's overall situation.

This includes indirect and long-term effects, as well as the avoidance of potential issues (which means working with risks).[2]

In the short run, some executives may agree to invest in data as "everybody is doing it," "it is the thing of the future," or "we should not fall behind." In the long run, however, particularly the CFO will want to see a calculation of the return.

The creation of convincing business cases, as outlined in Chapter 16, is not meant to make investments in data look sufficiently attractive so that funding can be secured. Instead, you need a commercial orientation from the very beginning. Determining a negative business case is not a failure!

---

[2] I had introduced the term "Return on Data" and its definition in my book *Becoming a data-driven organisation* (Treder, 2019).

# Collaboration

Collaboration is essential for several reasons:

- Every stakeholder gets involved.
- All areas of the organization are involved.
- The organization benefits from the joint creativity and knowledge of the entire workforce.
- It avoids parallel duplicate work (which leads to inefficiency and inconsistencies).

Unfortunately, collaboration does not happen automatically. The concept of collaboration in data matters is one of the first innovations a CDO should bring into an organization.

Collaboration requires an environment that enables and pushes it. This environment includes governance and organization, but it also requires incentivization and training. People need to understand why it is beneficial for them to leave their silos.

# The Data Office

Coordinated collaboration requires a powerful hub, without any own functional interests. By definition, the Data Office is such a place.

If you want to involve and leverage the entire workforce of your organization, this is my strong recommendation:

Set up a dedicated collaboration function within the Data Office, tasked with fostering and coordinating collaboration between all stakeholders.

# Clear data ownership

Without clear data ownership, everybody feels entitled to work with all kinds of data, while nobody feels accountable for the data. Such behavior leads to inconsistencies between silos.

Ownership makes people talk to each other. "I need X, and I know who owns it. That is why I am going to talk to that person" is the behavior you want to see.

# A decision and escalation process

You don't need an escalation process where people get along with each other very well?

I disagree.

If you leave the outcome of disagreement to personal relationships, it is likely that you won't take the best decision.

In contrast, a fact- and rule-based decision process assumes that

- All parties were acting with good intent, representing their respective functional position.

- The resulting conclusions could be unconsciously biased.

- But even fact-based functional conclusions could be suboptimal if limited to the impact on a single function.

---

### DATA MANAGEMENT THEOREM #3

It is not sufficient to base decisions on *some* facts. You need to consider *all* relevant facts – across the entire organization.

---

As part of such a decision process, "escalation" simply means taking the decision one hierarchy level up. You would ideally move it from a departmental perspective to a cross-functional perspective (which mitigates the bias) and eventually to an overall "shareholder" perspective (which resolves the bias).

Such a process fosters decisions free from personal perspectives and motives.

Furthermore, people even at lower levels of the organization are encouraged to develop cross-functionally optimal proposals right away, in anticipation of being asked to take a cross-functional perspective.

## Information sharing

You might want to share both official information and ideas. And this is not a one-way road. Everybody should be able to share their own ideas with others, ideally in a systematic way.

In essence, you need one or more technical platforms, but your governance also needs to declare it the only place to go. And the members of the Data Office can be expected to be the first to use the platform(s) actively.

## A Data Stewardship Network

How do you learn about different perspectives across all organizational areas? After all, any organization consists of functionally oriented people, geographically oriented people, cost-oriented people, customer-oriented people, and other varying shades of focus.

Yes, you need to shape a network to bring all of them together.

In addition to a collaboration platform for all employees, it is good to have a dedicated network for "data experts."

Such a network should be comprehensive enough to cover all areas of your organization, but it should be small enough to allow for joint conference calls and potentially other joint "real-time" activities. A promising approach is, therefore, to work with appointed Data Stewards, each representing a particular community.

A key idea of a Data Stewardship Network is to turn the flow of information from "batch mode" to "dialogue mode." In other words, discussions should become possible. Active moderation by Data Office staff would ensure focus, and documentation would allow for adequate follow-up.

Remember Data Management Theorem #2: Data Management is everybody's job – in every organization.

## Checklist

Your approach should...

- Link different departments together
- Link business and IT together
- Link all steps of the data supply chain together
- Reward collaboration
- Make people collaborate voluntarily
- Make it easy to work together
- Leverage synergies
- Avoid duplication and inconsistencies
- Define clear go-to roles and persons for all topics

## Motivation

I mentioned earlier that Data Management never works without a strong Board mandate.

Equally well, a strong mandate alone is not sufficient.

People who are forced to follow but are not convinced may find ways to boycott your work.

---

**DATA MANAGEMENT THEOREM #4**

---

A CDO requires both a Board mandate and the buy-in of the employees.

The former must be there from the beginning.

The latter must be achieved by the CDO.

---

# Cross-functionality

The *rule of optimization* can be derived from an old Operations Research discovery:

---

*The best data approach for the entire organization*

will always be better than

the sum of all parts' best data approaches.

---

In other words: If you optimize different parts of the organization independently of each other, you will not achieve the best setup for the organization.

Why is that?

Think of cases where you can indeed work on different areas independently, such as a graphics accelerators, linear equation systems, or MapReduce. What is a fundamental precondition for such approaches to work?

It is the absence of **interdependencies**!

Two parallel processes are interdependent if a final or interim result of one process influences the outcome of the other process.

We face the same situation with data in any organization: It reaches across functional and geographic borders.

Examples? Payment morale of customers, as recognized by Finance, should be shared with Sales. Customer information gathered by sales representatives should be made available to Customer Service.

It is for good reasons that people increasingly talk about "Customer 360": All aspects of a customer, usually gathered by various departments, are to be consolidated and made available to all potential users within the organization. The 360 degrees of a full cycle are a good analogy for the underlying idea: the customer is looked at from all angles.

This approach is not simply improving the activities of various functions. Important development such as omnichannel solutions even for B2B customers is simply not possible if customer data is not made available systematically across all channels.

Such a 360-degree perspective, however, is not limited to customer data. Product data is relevant for Sales, eCommerce, Marketing, Production, Finance, Customer Service, and so on. Market insights are relevant for Pricing, for Product Development, for Marketing, and for R&D, to name just a few.

From a data perspective, it is well known that the sum of locally optimized solutions is usually different from an overall optimized solution. As mentioned before, the reason for this is the number of interdependencies between the different areas.

The following two examples should make it easier to understand this mathematical rule:

- Sales want to increase revenue which is easiest if they agree on discounts. Finance, however, sees discounts as a minimizer of margin.

- Projects want to finish on time and within budget. This may lead to suboptimal solutions as it doesn't take into consideration the impact on future projects and other departments.

This is an organisational challenge: Team members often know what would be best for the organization – but they are afraid of their boss who is incentivized (formally and/or informally) by the isolated results of their own area of responsibility.

We need to be aware of the fact that there are no shareholders of any single department. Creating value for one department does not make sense from an overall perspective if this leads to higher costs across the other departments.

That is why it is imperative to consciously address this situation at the Board level:

**By making everybody focus on the benefit of the entire organization, all the time.**

An organization can achieve this target by applying the following principles.

# Focus on organization-wide targets, not on departmental targets

Both managers and individual contributors should be encouraged to do what they would do if they owned the organization. You would achieve this by aligning each individual's targets with the targets of the organization.

Where this is not addressed consciously, an organization will end up relying on people accidentally having the "right attitude." This approach will work for some people, but it will never work for everybody.

Management often talks about being proud of their "exceptional workforce that sets us apart from the competition." Let's be honest: It would be an implausible coincidence if all the good people within one industry decided to work for your organization, while all the bad ones joined your competitors.

Instead, you would motivate and incentivize people in a way that even the most selfish employees voluntarily do the best for the organization – because it is the best for them personally as well![3]

And you will regularly need to assess what motivates people to act – it changes over time.

## Incentivize collaboration

Have collaborative targets – incentivize the creation of value outside an employee's own area.

Make it part of business case structures as well! Projects need to get more resources, time, and money if they create solutions that are best from the organization's perspective, that is, where other departments benefit as well.

It also helps to ensure people from different departments know (and trust) each other personally. For this reason, get-together events and organization-wide celebrations of successes create very tangible value.

## Make the focus part of your organization culture

You should foster a culture of "Putting yourself in the shoes of the organization owner" (even if there is no single owner).

This culture is not identical to "focus on financial shareholder value!" Each organization has to focus on the triangle of shareholders, employees, and customers. A wise owner will do precisely this. What many organizations are lacking is a culture of encouraging **every single employee** to take the same perspective.

---

[3]This applies to the entire workforce, from the laborer to the expert to the leader.

You might also want to collect shining examples of "organization-wide thinking" and to acknowledge the protagonists publicly. Such examples should not be limited to cases of people working well together. You could also think of cases where people unilaterally do something that other departments benefit from.

It would be counterproductive to consider some teams more important than others – even if an organization defines its success primarily as the outcome of a particular team, for example, the Engineering team. Instead, people should see an organization similar to the human body:

- A football player would perhaps consider his feet his most important body parts – but what would the feet do in the case of a torn ACL? This anterior cruciate ligament basically keeps your knee together, and a dysfunctional knee renders the best foot useless.

- What would the brain do without a strong body to execute?

- Furthermore, consider all the "support functions" of a body. Your large intestine is undoubtedly not the shiniest part of your body – but, man, do people suffer where it doesn't do a proper job!

- And nobody would ask whether the heart or the lung is more important. Any organism inevitably fails if either of both doesn't work.

---

## DATA MANAGEMENT THEOREM #5

Data Management needs to be cross-functional – because data is cross-functional.

---

# Change Management

Hardly any organization is fully data-centric today. In other words, having a focus on data means a fundamental change to nearly all organizations.

Or, as Kirstie Speck, Vice President, Consumer Insight & Analytics at biotechnology firm Abcam, put it during a data conference: "Realise you are in the Change Management business! Algorithms are the easy part."

Most organizations have faced the need to change a couple of times during the past decades. This has led to the development of a discipline specialized in managing changes. You can assume the resulting recipes to be applicable to the huge "data change" as well.

| DATA MANAGEMENT THEOREM #6 |
| :--- |

Becoming data driven requires change across the entire organization.

# Data Literacy

I remember discussing with a huge corporation's product manager about the logic of products and services. He knew what he wanted, commercially, but he couldn't describe it. Yet he had found out that the current setup was not flexible enough.

When I asked how the existing product model got implemented in software solutions in the past, he said "I don't know. But I build on the outcome. Our IT folks did it all for me."

This should not be the target! Instead, business folks need to understand how to articulate their business needs. They usually have a structure but are not aware of it. They should.

Sometimes, when being asked the right questions, business managers may find out on their own that there is a logical flaw in their thinking – how would someone outside their department be able to find out?

Needless to say that the product structure I had discussed with that product manager was not only insufficiently flexible. It was also implemented inconsistently across the organization – different people had understood it differently. The business owner was not able to find out, let alone solve the issue.

Thank goodness, you can address such situations, by improving your organisaton's degree of data literacy:

# Help employees understand data

Instead of "outsourcing" data logic to IT (who, in turn, may not understand all the business motives), it seems more promising to help business folks understand base principles of data and express their desired logic in an unambiguous language.

It is crucial in this context that you don't aim at having all employees *learn* data, just as somebody *learns* vocabulary. The subject will change permanently anyway. Instead, train people to *think*. They should be able to make up their minds, and to follow any changes on their own.

# Share knowledge

I spoke at an AI conference in Dubai in 2019 where Omar Sultan Al Olama, Dubai's Minister of State for Artificial Intelligence, held the opening speech. He reminded the attendees of the fact that the Middle East used to be leading in science globally in premedieval times. Within 200 years, that advantage was lost. What was the turning point?

According to the Minister, it was a tiny invention by a man called Johannes Gutenberg: The mechanical movable type printing.

This invention brought books (and with them, knowledge) to the masses – except for the Middle East where books remained banned for centuries.

What has Dubai learned from this experience? They will not let the next revolution pass by again. That is why they are the only country with a dedicated Minister of State for Artificial Intelligence.

But what can **we** learn from this story?

Sharing of knowledge is vital if you want to capitalize on the power of your workforce.

Organizations should democratize knowledge by training everybody on data.

# Share data

Everybody should be provided with data to analyze. This is not limited to skilled data scientists – even blue-collar workers can draw valuable conclusions from data that describes their immediate environment. And everybody needs to be encouraged to speak up in case of findings based on data.

A positive example: "Since I changed the sequence of steps A and B three days ago, the number of defects has gone down, as I can see from the figures on the whiteboard. I will stick with the new sequence, and I will share my findings with my supervisor."

Even information security and data privacy constraints leave enough data available to the workforce for the entire organization to gain valuable additional insight: Aggregated or anonymizes data can reveal trends without requiring the credentials of individuals.

---

### DATA MANAGEMENT THEOREM #7

Data handling is *not* a topic for a small group of experts only.

Data-driven organizations need to upskill their entire workforce, and they need to provide them with access to all relevant data.

---

# The Data Supply Chain

Typical marketing analytics...

*"Yeah, but look at how many darts I've thrown!"*

**Figure 3-1.** Covering Analytics alone is like playing darts blindfoldedly

© Martin Treder 2020
M. Treder, *The Chief Data Officer Management Handbook*,
https://doi.org/10.1007/978-1-4842-6115-6_3

Data doesn't fall from the sky. It is captured or acquired, managed, stored, transformed, forwarded, and it finally gets used.

None of these steps should be looked at in isolation, as they strongly depend on each other. This is also why an organization's Data Office should be responsible for the entire lifecycle of data. I call it the "Data Supply Chain."

Let's have a look at the seven steps of this Data Supply Chain.

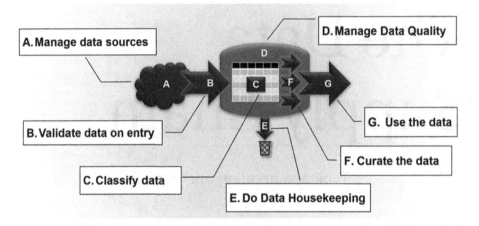

**Figure 3-2.** Seven steps of the data supply chain

# A. Manage data sources

Data is not just "there." A lot of it is consciously created by your organization, some of it is acquired, and sometimes it is provided by third parties. Even where it comes as a side product, it may still be of high value.

Typical sources of data are as diverse as

- Data entry by employees (via keyboard, voice, scanner, camera, or other devices)
- Incoming emails and messages
- Interpreted images of paper documents
- IoT devices
- Machines
- Software (e.g., log files)
- Data shared between organizations

- Social media

- Data found "somewhere" on the Web by data scientists

- Customer data entry on your website

- Streaming data (continuous flows)

- Both Masterdata and transactional data from external data providers

Not managing and not even coordinating these diverse data sources in their entirety leads to problems of which the four most severe ones are

(i) **Duplicate data acquisition**

Certain types of data are required by different functions. Without coordination, all of them may acquire the same data independently.

I recall an organization where the global head office alone had three independent contracts with Dun & Bradstreet. This is a waste of money. Managing the acquisition of data once for multiple users is undoubtedly cheaper.

Furthermore, sourcing data once reduces the overall effort. That is why this even makes sense for data that is available for free.

(ii) **Inconsistent data**

If teams search for data independently of each other, they usually end up using different sources, or they acquire the same data, but of different age.

All of this is undermining the "Single Source of Truth" principle, particularly for the kind of data that is out of control of your organization!

(iii) **Ambiguity of data**

The meaning of data found on the Internet can be ambiguous or unclear.

Such data, however, is a common source of information for Data Scientists. They often download data files with headers of which they need to interpret (to avoid the word "guess") the meaning. And even where their assumptions are correct, the precise definition of a column will be unclear.

(iv) **Risk of bias**

Where each team or department is responsible for the acquisition of required external data, neutrality may not be guaranteed. People may be tempted to select data that supports their objectives. Or they may work with multiple sources and finally go for the source that results in the most favorable outcome.

How about your own organization? Does anybody in your organization have an overview of all data sources? Are there any rules for the acquisition of external data, and are these rules enforced?

All of these activities are tasks of the Data Office.

But how do you set up such an organization?

- Establish a central function responsible for all external data sources.

- Make this function the go-to place for people searching for information.

- Ensure this function coordinates all activities and provides full transparency.

- Leave execution and ownership with expert departments. Ensure through Data Governance that nothing happens that is not known here.

- Task this function with the clarification of all definitions of external data.

- Have them match all incoming data with the corporate data model.[1] This model needs to tie external and internal data together. A typical example is the alignment of Dun & Bradstreet data with your internal customer database.

# B. Validate data on entry

When is the best time to validate your data? Yes, upon entry. It is easier to keep clean data clean than to revalidate data over and over again.

Not all data needs to have the highest possible quality – but you always need to know how good the data quality is.

---

[1]See also Chapter 18, section "The CDM and external data."

Furthermore, data does not only need to get validated and cleansed. It also needs to be

- Synchronized with internal data (including history)
- Anonymized/obfuscated where necessary
- Tagged with metadata

# C. Classify data

Data has a structure, and this structure needs to be understood by anyone who wants to work with it.

Most organizations have a structure in place when it comes to Reference Data. But it is essential to look at all data sources, to ensure they fit together despite their different origins.

A well-documented Data Model is critical for the documentation of how the business players want to run their business. At the same time, it needs to be good enough for IT to translate it into their database schemas and solutions.

IT Architects do not only need to understand the business data model to optimize the data structure for physical implementation. They also need to understand the data so that they can propose adequate hardware, platforms and solutions, the physical data model, the sizing, and the type of database.

---

## EXAMPLE

If a list of product features is straightforward, the IT Architect will go for a relational database. However, a complex product portfolio may lead to a more complex structure:

- Features may be optional, mandatory, or conditional.
- Some of them may have a variable number of parameters.
- Each of these parameters may come with different units of measure (kg/lb).

In such a case, a SQL database can become very complex, consisting of multiple tables. Furthermore, most of its fields would be blank. That is why a NoSQL database may be the better choice.

The final decision will also have to take into consideration *how* the data is going to be used. Inserting records into a SQL database is usually faster than doing so with a NoSQL database. You can see again that a lot of business input is required to decide on the best IT architecture.

---

Even unstructured data needs to be classified. No meaningful data source is too unstructured to determine and document its structure. You will at least require the corresponding metadata to understand the data and to be able to work with it. The level of required metadata depends on the content of the underlying data and its business purpose. All of this is nontechnical information to be gathered from business people.

**Summary**: Translating business knowledge and logic into sufficient information for IT Architects is a critical task of the Data Office. And unlike functional business representatives, the Data Office always has to take a cross-functional perspective.

# D. Manage data quality

In an interview with Information Age, Andy Joss[2] stated:

> *AI and machine learning are trendy. But if you are training an AI model, having really good data gives you more confidence that you will get the right outcome.* (Baxter, 2019)

This is a very polite translation of the more blunt statement "garbage in – garbage out."

If the data you are using is of low quality (or if at least you don't know whether it is good), you cannot trust in any outcome.[3]

Even worse, you often cannot tell from the outcome whether the underlying data was good or bad. This is different from food, where your palate or stomach will give you immediate feedback on the quality of a dish.

Imagine you calculate the chances of a product launch, and you obtain a result of "62.3 percent probability of success." Depending on the quality of your data, such a result can equally well be accurate or total nonsense – even if the calculation method is perfect.

Now imagine the product gets launched. No matter whether it becomes a success or not, you would not even know in hindsight how good the forecast was, as both failure and success are in line with its prediction.

---

[2] Andy Joss is Head of Data Governance – EMEA-LA at Informatica, located in Nottingham, UK
[3] For further aspects, including the impact of personal bias, see Chapter 10.

# E. Do data housekeeping

Lots of data may make you lose the real gems out of sight. It is like pouring good wine into a barrel of water. That is why you should install a permanent filter that checks which data to discard and which to keep.

Furthermore, data retention regulation needs to be actively managed. On one hand, you are often obliged to keep certain data records for a minimum amount of time. On the other hand, data privacy regulations, such as GDPR, do not allow you to keep everything as long as you might want.

Either way, failing here can get extremely expensive.

# F. Curate data

Having perfect data in your repositories is not sufficient for its usage.

You also need to ensure everybody uses the correct data, and that it is used sustainably.

## The purpose of data curation

Let's look at it from a data user perspective – you can use this as your checklist:

What does the user need to know?

- Where do I find the data I need?
- How do I use the data?
- What is the explanation of the data?
- Who owns the data (in case I need changes)?
- Has someone else started to use the data before? What can I reuse?

And what will a user need to rely on?

- The data provided to the user is correct, the single version of the truth, always up to date, in line with its description, and sufficiently highly available.
- Any issues and changes will be communicated well in advance.
- Data security and privacy are taken care of.

# Aspects of good data curation

Data needs to be used in a *sustainable* way. In other words, it is not sufficient for data to be good at the moment of its acquisition or its first usage. Whenever the content or structure of data changes in future, the new data needs to be used automatically, without any need for human intervention.

In concrete terms:

(i) **Centralize data storage**

Don't let people work on raw data – have them use web services to update or use it.

(ii) **Centralize data logic**

Don't implement general data logic in client applications – have it implemented centrally. Client applications should call a web service to apply the logic.

(iii) **Have users understand the data**

Ask the data user to talk to the data owner. The latter is the best person to explain the intended logic of the data.

(iv) **Give access to data**

Provide users with access to all data they need, together with information about its structure, interdependencies, quality, end-of-life, and any other relevant attribute.

This applies to all types of data: Masterdata, Metadata, and Transactional Data.

(v) **Understand and document data transformation**

Throughout the Data Supply Chain, data gets merged, filtered, calculated, derived, or enriched. Many such activities have been in place for decades, long before Data Management became a recognized discipline.

As a result, your Data Supply Chain has existed for a long time as well. In most cases, there is no systematic documentation of the logic behind those various kinds of data manipulation and transformation.

However, in order to manage your Data Supply Chain adequately, you need to understand what is going on.

In most cases, it is not effective to start an assessment on the business side, that is, to ask the business owners how data should be logically transformed. They usually don't have the expertise to be able to explain.

Instead, you would need to reverse-engineer the logic that past generations have implemented. This is a long and tedious process, including challenges such as multipage SQL operations and COBOL source code.

So, what are the benefits that justify such an endeavor?

The first (and most important) benefit is that you will obtain the full picture about what is currently being done to your organization's data. No matter how good the original data has been, if you don't understand how it was subsequently manipulated, any resulting data repository is useless.

Secondly, you can ask businesspeople to validate the outcome against their business concepts. It is easier for them to validate what you present to them than to specify it from scratch, on their own.

Thirdly, you may be able to reduce the complexity of these data transformations. If a business expert does not understand the data transformation logic, a simpler logic may work as well.

Finally, if businesspeople understand what is being done to "their" data, they will more willingly take ownership.

As you discover and document all data manipulation, you would put it under change control: Nothing gets changed or added without following a robust process, of which documentation must be an essential part.

## (vi) **Document data usage**

Giving everybody access does not mean you don't know who is using it: In case of changes to data structure or content, you need to know who is impacted right away.

Yes, it is a lot of additional work to always log who is using what. Initially, it may sound like overkill. But most organizations grow to a point where the knowledge does not fit into people's heads anymore. Even where it does, consider that people leave the organization or retire. Documentation is there to stay.

## Provision of information to users

Documentation and publication are essential: Data formats, interfaces, technical access description, data owners, and data change processes need to be transparent to any (potential) user. Please think of the following three components:

(i) **Data Catalog**

Data Catalogs provide users with all the relevant information about the data they need.

(ii) **Web service directory**

This is the technical description of data access for application developers and data scientists without which people may re-develop existing logic or keep tracing through source code to find an adequate web service.

(iii) **Intranet site**

People outside a dedicated data team may not know where to look as they cannot classify their needs. They may not even know the difference between a Data Catalog and a database. They should not have to!

Instead, you would offer them a single point of information where they find all the information in an easily digestible way.

Such an Intranet site would then refer users to a Data Catalog or a web service directory in a user-friendly way. It would also provide access to explanatory documents (including classic presentation files), data policies, processes, and so on.

Finally, users should find a feedback channel, where they can point the Data Office to missing or wrong information, and where they can provide proposals to improve the curation of data. This aspect should be part of your data workflow.

# G. Use data

Finally, we get to the step that many organizations decide to start with.

But even the usage of data is more than fancy data analytics and colorful charts.

The usage of data in an organization is as multifaceted as the data itself.

Typical usages are

- Operational usage
- Calculation of Key Performance Indicators (KPIs)
- Reports
- Analytics (AI, Data Science, ML)
- Robotic Process Automation (RPA)
- Audit support
- Usage for testing or validation purposes

All of these require consistent data. That is why none of the preceding steps should be dealt with by an entity that covers only one or two of the areas listed.

Data Analytics folks tend to prefer "fixing" the data before they use it. Don't support this desire; don't work backward from Analytics. You might render your data useless for other important purposes.

# Summary: Cover the entire data supply chain

I have seen too many organizations appoint a Chief *Analytics* Officer and hope for great insights. Such organizations limit their data focus to a subset of step G. They do not even cover all aspects of data usage, let alone the essential steps before.

Let's have another look at the initial diagram of this chapter. It becomes apparent that Analytics represents just a fraction of a Data Office's responsibilities.

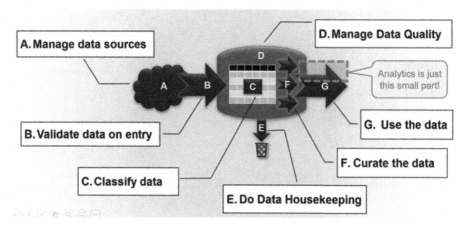

**Figure 3-3.** The Data Supply Chain contains more than Analytics

Why do executives often focus solely on Analytics and visualization? They consider only the *visible* parts of managing data!

This approach is similar to architecting a building that solely consists of a viewing platform high in the air. You don't need to be an architect to understand that gravity will prevent this approach from being successful.

Using a helicopter instead will indeed provide you with the desired view – but only temporarily, until you run out of fuel.

The usage of data follows the same logic. That is why covering the entire data supply chain will always pay off. Lay the foundation, then erect the building floor by floor. When you finally arrive at the viewing platform, you can be sure it is stable and sustainable.

## DATA MANAGEMENT THEOREM #8

Data Management needs to cover all steps of the data supply chain, from the creation or acquisition of data up to its usage.

# Data Vision, Mission, and Strategy

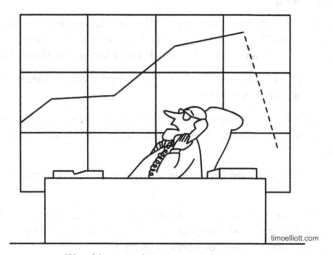

*"Yes, I have made a strategic decision.*
*I've decided to ignore the bad news... "*

**Figure 4-1.** Ignoring data is a strategy as well

© Martin Treder 2020
M. Treder, *The Chief Data Officer Management Handbook,*
https://doi.org/10.1007/978-1-4842-6115-6_4

# Data strategy – seriously?

## Strategy vs. execution

Successful business leaders often stress the need to execute. Ken Allen once said: "Most strategies are like New Year's resolutions, full of really good intent but no desire or capability to execute" (Allen, 2019).

In other words, it is essential to focus on execution. However, running fast is not sufficient. You need to know where you are heading.

Ken Allen also said "At DHL I needed a clear turnaround strategy that everyone would be able to execute. I needed to make sure that the whole company was focused on 'doing things' – the right things" (Allen, 2019).

This quote makes it clear that a strategy is required – as long as it guides everybody in the organization and leads to action. All aspects of a strategy that do not serve these purposes are a waste of energy.

## Strategy in times of Agile

Isn't an explicit vision, mission, or strategy too static, given the rapid changes we are observing? Isn't all of it outdated before it is executed? Isn't Agile replacing all planning?

Indeed, Helmuth von Moltke[1] was right 150 years ago when he said: "No plan of operations extends with any certainty beyond the first contact with the main hostile force" (Wikiquote – Helmuth von Moltke the Elder, 1871).

It is noteworthy that even von Moltke always had a plan – but he was ready to change it whenever necessary.

The main reason for the need to set the direction and to plan forward is that people always need guidance – no matter how quickly that direction may change again.

You also require a plan and a baseline. Otherwise, changes quickly become arbitrary.

Finally, the current direction, plan, and baseline need to be made visible to all stakeholders. How would you be able to communicate any change of direction if the previous direction was not communicated?

---

[1]Field Marshal Helmuth Karl Bernhard Graf von Moltke (October 26, 1800–April 24, 1891) revolutionized army orchestration during wartime when he was Chief of Staff of the Prussian General Staff from 1857 to 1871.

| ANALOGY |
| --- |

Imagine a situation where several tugboats try to jointly pull a big vessel through a narrow harbor by means of towlines.

To be efficient, all tugboats have to pull in the same direction.

And no matter how frequently the vessel requires a change in direction, all tugboats have to execute every move synchronously.

This requires clear communication of the direction and of every necessary change.

In organizations, the overall direction is described and shared through Vision, Mission, and Strategy.

Critical, cross-functional topics such as data should follow the same path, as the same reasons apply.

# Culture eats strategy for breakfast?

True. However, culture would starve to death without having a good strategy for breakfast...

But what makes a strategy "good"? A good strategy serves a purpose: It drives behavior.

A CDO needs to ask the right questions before developing a vision, a mission, and a strategy for data in the organization:

- Vision: What would this organization's way of managing data look like in, say, five years from now (the way things are at the moment)?

- Mission: Which targets do we intend to achieve to meet our vision?

- Strategy: What are business priorities? Which steps do we take to get there?

After you have developed a vision, a mission, and a strategy, you should be ready for the fourth step: Developing your plan based on your strategy.

A plan should always cover activities supporting your strategy as well as tactical or operational activities. A complete plan helps you prioritize your activities.

Don't get hung up with difficulties in categorizing your thoughts as strategy, vision, or mission. As long as you achieve the purpose behind the different categories, you are fine.

Whatever you might have learned about any of these categories in general, feel free to apply it to data as well.

---

## ANALOGY

Think of mountaineers.

- Your *vision* is the selection of the peak you want to climb and the date until you want to have achieved this performance.

- Your *mission* is to do it during a certain time of year, within a certain amount of time, and with a certain performance per day. You can check every evening whether your mission is on track.

- Your *strategy* consists of realization aspects such as the number of team members or the equipment.

- Your *plan* describes the daily execution. It may be subject to change, whenever preconditions change.

---

How does all of this relate to the overall direction of your organization? Have a look at Figure 4-2.

**Figure 4-2.** Corporate direction and data direction

You should base your data vision on the corporate vision. At the same time, your ideas may influence that overall vision, for example, by making "becoming a data-driven organization" part of the corporate vision.

It must become apparent that your vision is a considerable part of the overall vision of your organization. James Wilson from Gartner Group stressed during a Gartner conference: "This is not about a small Data Management team delivering small pieces. It is about the company changing their thinking and rethinking their business."

For this to become concrete, your data mission should describe which components you use to reach your data vision. This is a Data Office–specific exercise.

Data strategy should again be influenced by both your data mission ("**How** do I proceed with all selected components?") and the corporate strategy. Again, your data strategy may influence that corporate strategy as well, for example, in setting your organization's long-term priorities.

# Vision

## What should a vision accomplish?

The primary purpose of a data vision is to ensure all stakeholders are on the same page regarding your organization's data ambitions – particularly the Board and you.

Your vision should answer fundamental directional questions in three aspects:

a) What is the role of data in relation to your business model?

- Is data expected to support the **existing** business model?

- Is data expected to improve those business models, for example, through digitalization?

- Do you want it to create **new** business models?

- Is data supposed to become a business model **on its own**?

b) What does your organization want to achieve through data?

- Do you want all customers, partners, and stakeholders to **trust in your organization's data** as a reliable basis for business decisions?

- Do you focus on the valuation of **data as an asset** to become part of your organization's culture?

- Is established **data governance** your primary vision, assuming that this allows you to achieve all the other data-related targets?

c) What is the data handling of your organization supposed to look like in a few years from now?

- Do you intend to have your data office support your organization's business in the background, or do you rather want to actively co-shape the business model?

- Do you want to build a strong data management organization, or do you want to focus primarily on strong data competencies within the business functions?

- Do you intend to create a strong foundation for *all* usages of data, or do you want to focus on best-in-class Data Analytics capabilities?

# What should a vision focus on?

You might not need to take all the preceding points into your vision. Fewer points may make each point easier to remember (and to remain in focus).

It is helpful to consider what you need to tell your organization. Do they need to hear the entire list? Or are some points clear anyway, and instead you need to teach the organization the importance of those points that are new to them?

It is definitely a good idea to drop any well-understood parts (or sum them up in one bullet point) so that people focus on the new aspects. Doing so doesn't render the other aspects irrelevant.

In some cases, you may want to focus on specific areas of Data Management, where you see great opportunities or a big gap. Your vision is an excellent place to focus the organisation's data aspiration on these points.

In case your organization tends to be very conservative, your vision could stress the target to become an open-minded organization in data matters. Or, as Isabel Barroso-Gomez summarized it when she was Head of Data Governance at Sparebank 1 in 2018, "add curiosity to your vision."

In case your organization used to limit data management to a few topics, for example, Masterdata Management or Reporting, it may be valuable to stress that various different aspects of dealing with data are equally important.

An example could be an intended balance between today's colorful Analytics reports and as yet neglected tedious Data Quality work.

To indicate this balance, you might have your vision consist of the targeted number of pillars or focus areas.

Why not use a chart to illustrate where you see the organization today and how quickly you intend to develop it toward your data vision?

The example in Figure 4-3 bases on the three pillars **Fix**, **Optimise**, and **Innovate**. You could add organization-specific aspects to each of the three pillars below the bars.

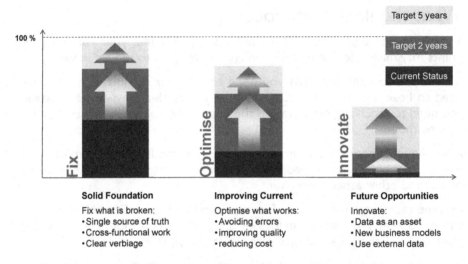

**Figure 4-3.** Vision example – Fix, Optimise, Innovate

In case you want to stress similar relevance of exactly three areas, you might illustrate this using a three-dimensional coordinate system, as shown in Figure 4-4.

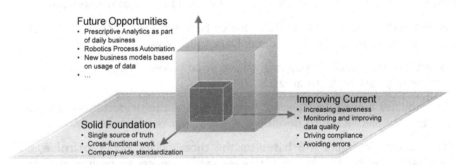

**Figure 4-4.** Vision in three dimensions

Alternatively, you can formulate your vision with a focus on the target state of your organization with regard to data, for example:

*By (year), we want to become an organisation where…*

*… data is centrally governed and standardised*

*… data is understood by business functions, not only by IT*

*… data is becoming the common language between all business functions*

*… financially, we treat data as an asset, at least internally.*

And, remember, a brief vision requires a lot of preparatory work with all stakeholders, so that it becomes meaningful and relevant to everybody's daily work.

# Mission

As you have seen, a good vision is indispensable. However, it may be a bit too abstract to directly derive your strategy from it, let alone a concrete action plan. This is where a clear mission can help. A mission describes concrete targets your organisation wants (and empowers) you to pursue.

Critical aspects of an effective mission statement are

- Your mission should be measurable so that you can regularly check your progress. A good mission statement will help you determine gaps or delays over time.

- Your mission should consist of a comprehensive list of concrete targets. Each of these targets would support your data vision – and thus, the vision of your organization.

- Furthermore, good targets are concrete (and concise) enough to obtain Board endorsement – which you require before you detail out your implementation plans.

There is no one-size-fits-all data mission, of course. But the following **seven aspects** of an organization's Data Mission could be a good starting point.

I invite you to translate this list into your specific Data Mission, in the context of your organization's gaps and opportunities. You also find a few comments on each of the aspects in Figure 4-5.

**Figure 4-5.** Data Mission example

## Define centrally governed data handling standards

This target would expect the Data Office to introduce data processes, a single data glossary, data policies, and rules. You would not execute everything centrally, but you need to set up organization-wide standards.

Developing such standards together with all relevant stakeholders will allow a Data Office to become the "data glue" and the translator between all functions across the business and IT.

## Introduce cross-functional MDM, based on a single source of truth

While you might not need to centralize the maintenance of Masterdata and Reference Data, you'd certainly want to stop independent activities in different areas. You can achieve this through a centralized governance, a single set of Masterdata repositories, and well-defined Masterdata consumption standards, for example through web services.

A resulting implementation plan would require a long-time horizon. Most probably, this is the area with most skeletons in your organization's closet. Think, for instance, of hard-coded Reference Data or local, ungoverned maintenance of Masterdata.

# Ensure good Data Quality through measurement and improvement initiatives

Again, you wouldn't want anybody from the Data Office to become responsible for fixing Data Quality issues. Instead, the owners of the data need to be in charge. But they have to know how to do it, and they should do it based on common standards, allowing for data quality to be compared across the organisation.

This target allows you to set up an organization-wide way of measuring data quality, of finding root causes of issues, and of determining the responsibilities of addressing those root causes. The Data Office may be the facilitator and rigorous guardian while the business functions execute.

And you don't need to reinvent the wheel; Six Sigma is there to help – both you and the business functions.

# Work with business people to turn data into information

The Data Office teams will work with people from all business functions to turn data into information. They will use modern methods of Data Analytics, from Reporting to AI.

As with previous points, the Data Office may not need to do all the Analytics and Data Science in a centralized ivory tower. Instead, you might wish to find a balance between centralization and federation of responsibilities.

This target would empower you to work with business folks and IT on the best possible setup. You would aim at finding the right balance between business proximity, collaboration, and synergies when organizing the work of your organization's Analytics experts.

# Have all of this supported by the right toolset

A Data Office should not create a Shadow IT department, nor should any business function feel inclined to do so. However, a Data Office might be well suited to team up with IT to serve your organization's business functions. You could start with a delineation process between the Data Office and the Information Technology teams, potentially led by a neutral moderator.

You would also need to listen to the business folks to understand their priorities. (I can tell you, for instance, that they hate lengthy, bureaucratic processes.) As you are also responsible for the long-term perspective, you will need to find a balance between sustainability and agile methods.

## Implement adequate ethical standards in dealing with data

A Data Office needs to find a balance between what is possible, what is allowed, and what is beneficial. To do so, you would usually have two guiding forces toward ethically strong data handling: The law and the well-being of humans (particularly your organization's employees and customers). None of the two alone is sufficient – neither formal compliance while misusing gaps in the law nor a well-meant people-friendly approach that doesn't consider formalities of the law.

This target requires both a big program and an ongoing effort.

## Train and connect all entities in data matters

Not all financial topics need to be solved by people in the Finance department, just as human well-being is not the responsibility of the HR department alone. Similarly, data is an organization-wide task, not just the responsibility of one central Data Office.

This target would ask you to set up organization-wide data networks to link people across functional or geographical borders and to establish proper training and communication to shape a data-literate workforce.

# Strategy
## Why do you need a data strategy?

A good strategy allows teams to take left-or-right decisions and to prioritize, as a critical factor in addition to business cases, resources, dependencies, and so on. It helps avoid the need for fundamental discussions in each single case.

Having such directional support is particularly important where you face different equally valid options to proceed. It also ensures consistency across different decisions.

A typical example is a decision between attracting customers through perfect solutions and attracting customers through competitive prices. Both options are valid, but mixing them leads to blurred brand recognition.

## How is the strategy positioned in comparison to the strategy of the organization?

Your data strategy is not a strategy on its own. The strategy of the organization must be in the lead.

As the internationally renowned author and strategic advisor Bernard Marr once put it, "Instead of starting with the data itself, every business should start with strategy" (Marr, 2019).

The example above shows nicely how the data strategy relates to the corporate strategy:

- On one side, data management needs to support brand recognition, for example, by stating "We want to become known as the lowest-cost competitor among all high-quality providers."

- On the other side, being a low-cost provider does not mean taking a low-cost data approach. An organization can decide to invest heavily in data to find out how to be the provider with the lowest prices while staying profitable.

That is also why there is not one perfect data strategy. It heavily depends on the organization's overall strategic direction.

How would you develop your data strategy? Ideally, in close alignment during the development of your organization's strategy. A CDO should be involved in the development of a corporate strategy. There may be reasons to influence the latter, based on the knowledge of what is possible with data.

What if you join an organization as a CDO which already has a corporate strategy? Accept it, and match it. Describe how data management can support that strategy. Be ready to contribute to any further development of that strategy.

## How do you develop and maintain your data strategy?

Foresee regular strategy matching reviews. Ask: Are we still on track? If not, is it because the strategy points us in the wrong direction, or are we not realizing it properly?

Should you find out at any time that your data strategy does not adequately support your corporate strategy (anymore), start a process to adapt it. Don't wait for the next review of the corporate strategy.

At the same time, feel free to provide feedback from a data perspective into the overall strategy at any time. Just don't forget that data is not the driver of that corporate strategy.

## Your individual measure of success

We have been talking about the vision, mission, and strategy of a Data Office as well as their relation to their corporate equivalents.

But how do you gauge the success of your work in the course of time?

There may be some aspects you would not put into official statements but share with your team and review regularly.

A typical internal success criterion is the *standing* of the Data Office and the CDO within the organization. While being a precondition for success, this criterion would usually not make it into any of the published versions of Vision, Mission, and Strategy.

Particularly in the early days of a Data Office, you cannot rely on any tradition. You have to earn all the respect your counterparts in Legal, Risk, or Security may already have inherited from their predecessors.

Progress in this area is hard to measure. It is still worthwhile thinking about it to avoid self-delusion.

A Data Office can be considered one of the critical support functions that need both the arguments and the authority to make people change their behavior. So why not use other departments in a similar position as your benchmark?

A somewhat measurable criterion could be "How often do we get invited to support others?" It can document the transition from "being imposed on others" to "becoming a trustworthy advisor."

This aspect is so vital that you might want to add it to the individual targets of your team members.

Here is another test of how successful your measures are: Listen to employees, and find out what they mean when they say "we" in the context of handling data. Are they referring to the entire organization, or is it their respective microcosmos, for example, their team or their department?

I recall a Senior Vice President of Finance who said: "**We** need to work cross-functionally." Unfortunately, in this sentence, the word "we" stood for the Finance department. The Accounts Receivables team was expected to work closely with their Controlling colleagues, while a dialogue with, say, the Sales department was not considered a priority.

In mature organizations, "we" is mostly used to describe that entire organization. This is tremendously important in Data Management as there is no such thing as "Finance data" or "Marketing data." Instead, we should talk about a Finance or Marketing perspective of "our" organization's data.

Similarly, the word "they" should not be used to describe other teams (or, in the sense of "those folks on the executive floor," the top management). It should instead refer to parties outside the organization, predominantly the competition.

# Masterdata Management

*"You want my metadata?
You'll have to pry it from
my cold, dead hands..."*

Figure 5-1. Who owns the data?

© Martin Treder 2020
M. Treder, *The Chief Data Officer Management Handbook*,
https://doi.org/10.1007/978-1-4842-6115-6_5

# Isn't static data old-fashioned?

Data Management is about Analytics, Machine Learning, and insight through visualization, right? So, why does this book deal with Masterdata right after its Strategy part, even before covering Data Governance?

And, yes, in times of Agile, Masterdata sometimes sounds like a constraint from the past.

Nothing could be further from the truth.

In a LinkedIn post in December 2019, the "Data Whisperer" Scott Taylor, put it this way: "Master Data is Macro-Trend Agnostic!"

And I wholeheartedly agree. While changes to Masterdata may need to become more and more agile, the need for actively managed Masterdata is not going to disappear any time soon.

Transactional data reflects daily transactions or events, often even created as a side product. In contrast, Masterdata is maintained consciously to support business activities.

Even if Analytics concentrates on transactional data (including Big Data), a closer look reveals that none of that transactional data could be classified and interpreted (and thus provide business insight) without Masterdata.

Furthermore, transactional data created based on bad Masterdata will inevitably inherit the insufficient quality of the latter.

And this can become an expensive problem. I recall a company where the Head of Sales pointed me to a pile of letters on his desk, all of them addressed to customers and marked "return to sender – undeliverable." Services had been rendered for these customers, but the corresponding invoices could not reach them. The company's address data was too bad, and the financial impact directly hit the bottom line!

Of course, Masterdata is not limited to customer data but relevant to all areas of an organization. I don't know of any business function that does *not* need or own Masterdata.

And digitalization even increases the impact of bad Masterdata on all parts of an organization.

In 2019, Deloitte had asked Chief Procurement Officers about their biggest challenges in mastering digital complexity. The most frequently mentioned topic, brought up by a whopping 60 percent of all CPOs, was "poor Master

Data Quality, standardisation, and governance"[1] (Delesalle & Van Wesemael, 2019). Welcome to a world where a Chief Data Officer has become indispensable!

# What does Masterdata cover?
## Masterdata, Reference Data, Metadata

The delineation between Masterdata, Reference Data, and Metadata is not entirely unambiguous.

However, my recommendation is not to spend too much energy on finding a perfect definition. You need a good reason to distinguish any two types of data. Otherwise, it is a purely academic exercise.

A country's postcode format could be seen as Masterdata, but it could also be considered Metadata. An exchange rate could be regarded as Reference Data – as it is subject to permanent change, you could consider it transactional data as well.

Do you have to decide on a classification? Only if you have practical reasons to do so!

Let me offer you three potential drivers that would justify a classification:

- The need for different processes
- The need for a change in ownership
- A difference in the impact of changes

This will allow you to define governance for different types of data by referring to Masterdata, Reference Data, or Metadata.

Considering those three aspects, two criteria come to mind:

(i) **Different processes**

You would eventually want to distinguish between **changes to data structure** and **changes to data content**.

The former often enforce a redesign of applications, interfaces or databases. They are fundamentally different from the latter that ideally make applications behave differently through configuration. Changes to data structure usually require a different, more complex process than changes to data content.

---

[1] Forty percent of all CPOs mentioned "Inability to generate analytics and insights across these systems." No other point was mentioned by more than 33 percent of all CPOs.

#### (ii) **Different accountability**

You will find that **accountability** differs along those lines as well: Data about customers, locations, and products is defined by the business folks as part of their daily work, as getting it wrong in one case has a limited impact. Any change to data that structures or specifies such data has a potential impact on many records. The same applies to mass data changes, usually applied in batch mode. That is why the responsibility for such changes should be with the Data Owner who would sometimes even have to organize an impact analysis.

As a result, we could suggest the following classification:

#### (i) **Masterdata**

- Is maintained by people within the business functions.

- Individual changes have low if no impact on business processes.

- If a single record is wrong, the impact is often limited to transactions dealing with this record.

- Systematic flaws in maintenance processes can have a significant negative impact, though.

#### (ii) **Reference Data**

- Is accessed by IT applications that are used to maintain Masterdata.

- Changes to Reference Data may impact Masterdata. If a job title is changed or retired, all employees with that title are affected.

- Is maintained by data creators.

- It usually consists of sets of values from which attributes of Masterdata entities are selected.

#### (iii) **Metadata**

- Specifies reference data, Masterdata, and documents

- Is managed by data owners

#### (iv) **Transactional data**

- Is created through the execution of business processes

If you choose these criteria as your differentiator, you can more easily define different change processes for data that solely requires changes in a data maintenance tool vis-à-vis those that require an assessment of (and potential changes to) an organization's application landscape.

And again, your definitions should be driven by business purpose. If you intend to treat two different types of Reference Data differently, feel free to break down further. And if you apply exactly the same processes to Masterdata and Reference Data, use one term to describe both of them together.

But be future-proof: Keep the flexibility to distinguish in the future what is similar today!

Note that, throughout this book, I am additionally using the expression "Masterdata" as a generic term covering Masterdata, Reference Data, and Metadata. Accordingly, whenever I say "MDM" (for Masterdata Management), I refer to the management of all three types of data. You don't need to follow this habit, but you should be aware while reading.

# Examples of Masterdata

- Parties: Customers, suppliers, external partners, employees, authorities

- Service-related things: Products, spare parts, supplies, material

- Utilities: Forklift trucks, machines, vehicles, printers

- Geographies: Plants, warehouses, offices, sales areas, addresses

# Examples of Reference Data

The following examples demonstrate the multifaceted nature of Reference Data. They make clear that coordinating Reference Data and agreeing on rules is a central task, ideally to be assumed by the Data Office.

(i) **Exchange rates**

Every internationally active organization requires accurate exchange rates for transactions and valuations.

There are various different ways of calculating exchange rates, and you'd have to decide on the location and whether it is the selling or buying rate. It is equally vital to agree for every transaction in which currency the base price is calculated.

Furthermore, it is necessary to decide on the frequency of changes to an exchange rate - ranging from real-time changes to monthly rates, depending on the purpose.

If you forget just one single aspect, you will run into ambiguity. This leads to two different prices for the same purchase – a worst-case scenario both for your customer or supplier relationship and your Finance calculations.

Exchange rates change by the second, being subject to permanent trade. However, considering every change would make it impossible to work with a stable foundation, and prices for exports may change permanently.

Volatile currencies require quick reactions to exchange rate changes, and still, you would not necessarily sell the foreign currency within seconds after your purchase, leaving you with the risk of post-sales currency devaluation, no matter how precise your exchange rate processing is.

Exchange rates for transactions between business partners can usually be agreed upon freely – yet there is no room for ambiguity.

Regulated calculations which you face, among others, in the areas of taxes, duties, and reporting often have to follow external rules, including the frequency of changes and rounding rules.

In both cases, it is tremendously important to store the history of exchange rates. Otherwise, you would not be able to retrospectively calculate the applicable rate for a deal in the past.

All of these aspects demonstrate how important centralized handling of exchange rates is for any organization. If, say, Sales and Finance work independently, you will face severe reconciliation issues, and if you leave the choice of the exchange rate to BI and Analytics folks, different people will come to different results.

My recommendation is

- Carefully define and select a data business owner for this area. That business owner needs to align with all other stakeholders.

- Have the Data Office become the mandatory custodian, to avoid independent activities.

- Set up and document clear rules with the data business owner and the stakeholder community.

- Ensure you subscribe2 to all agreed sources of exchange rates and that you store the values in your Reference Data repository. It is paramount to always have an accurate and complete history of exchange rates for each source.

- Consider storing own, organization-specific attributes, for example, whether you accept a specific currency for payment or as of which level of volatility you send an alert to Risk Management.

(ii) **Address data**

Address information is complex. Both the addresses themselves and the allocation of addresses to business parties are subject to permanent change.

Even countries with regulated address logic face ambiguity.

- Typical parts of an address can usually be abbreviated which leads to different valid spellings of the same address (e.g., "rd" vs. "road").

- Some parts of an address are optional, that is, only provided for clarity reason and not necessary to determine the exact location.

- Many countries have different valid ways of determining an address, often because a new method was introduced, but people keep using the old one.

- Some countries have a sophisticated address and postcode system, but hardly anybody uses it.

- Different alphabets make it even more complicated.

All of these points create ambiguity, so that simple string comparison is not sufficient to determine whether two addresses specify the same location.

The entire topic of address management deserves a dedicated book.

---

[2]A typical example for an external reference exchange rate that you could subscribe to and make it the basis of your online price calculation is the "hourly closing rate," published by The World Markets Company PLC, Edinburgh ("WMR").

Recommendations:

- Normalize: Agree on a single leading standard and unambiguous mapping rules (e.g., a clear standard for transliteration in case of multiple alphabets involved).

- Consider using geo-coordinates: They are independent of spelling and format as they unambiguously specify a physical location. But be careful: Skyscrapers may come with hundreds of organizations having the same x and y coordinates. And the shape of a building does usually not correspond with the rectangular shape of a pair of geo-coordinates. Consider working with geo-fencing and using a third coordinate for vertical distinction.

- Consider external address cleansing services: No organization has perfect address data for all countries on earth, but some providers work with different local address providers to provide concise address data globally.

- For compliance and legal reasons, keep a history of addresses per party, and keep any original address as provided by the customer.

- Use country-specific address logic, including related Metadata. You can implement it yourself; use APIs or external data providers. In any case, consider the possibility of daily changes.

(iii) **Language data**

A language is a language? Of course, the world is more complicated than that...

Typical challenges comprise

- Different alphabets (Latin, Cyrillic, Hebrew, Korean, Chinese) most of which come with multiple subsets.

- Different spelling standards of the same language in the same country (Norwegian)

- Transliteration not always unambiguous (Kanji in Japan)

- Multiple country versions of the same language (English)

- Multiple languages per country (India, Switzerland, Canada)

Recommendations:

- Always work with Unicode when dealing with various languages.

- Define an internal leading language and alphabet. Always map to your default language. This gives you a mere (n-1) relationships to manage, instead of n*(n-1)/2 relationships between every two pairs of languages.

- Work with the combination of country plus language (as most language service providers do), and always define a default language per country.

- Language handling must be able to define specific expressions within your organization that must not be translated, for example, as they are part of the brand. Consider that you will still need to transliterate in most cases. And the handling of organization-wide translations of all other organization-specific expressions must be possible as well.

## (iv) Organizational hierarchies

Does your organization reorganize from time to time? Does it acquire and integrate other organizations? Does it enter new markets, with its own subsidiaries? Does it found joint ventures with other organizations? Unless you work for a local chimney-sweep, you will have answered "yes" at least once.

But who is responsible in each of these cases? Where is the structure stored and how?

Traditionally, people outside Finance tend to think that the organizational structure is intuitively clear. But the devil is in the detail, and there are many good reasons to avoid the frequently observed hard-coding of hierarchies in functional software.

Recommendations:

- Clarify potentially different purposes for organizational hierarchies. The internal structure of organizations is often different from the external, legal structure. If necessary, maintain two or more different hierarchies.

- Insist on the mandate for the custodianship of this information to be with the Data Office, to ensure sustainable, cross-functional, and comprehensive management of hierarchy information.

- Determine the best-suited data business owner. Depending on the profile of your organization, the owner may be Finance, Legal, M&A, or even a dedicated Business Transformation unit.

- Design the data structure of your hierarchies in a way that allows for changes to be implemented through configuration.

- Work with the business owners on the (nontrivial) mapping between different evolutionary steps of the hierarchy for an appropriate year-on-year comparison. Such a mapping must be made available to all data consumers to ensure consistent reporting and Analytics.

- The structure of the stored history data should allow for easy allocation of events and transactions to the right part of the hierarchy as soon as the date is known, particularly for legally oriented hierarchies.

# Examples of Metadata

There are as many types of Metadata as there are types of other data. Eventually, Metadata describes other data.

Here are a few examples:

### (i) Format descriptions

Most elements of Masterdata and transactional data require format descriptions, to allow for data validation and, if necessary, to interpret the data. Typical examples are the length and format of account numbers.

Format description of external data falls under Metadata as well, such as the format and length of VAT numbers.

### (ii) Document information

Classification and tagging of documents: Stored original documents have a huge number of metadata elements such as format, number of pages, or date of last modification. Most file formats even store a portion of the Metadata as part of the file itself.

Imaged documents have far fewer metadata tags such as origin, author, or size – here, additional information comes through OCR and text analysis. This is why these steps should take place upon receipt of any such image, to allow for its immediate findability and effective utilization.

Full OCR and text analysis often translate entire documents into systematic information, turning the original document image into a side aspect. In this case, most of the information can be considered transactional data, not Metadata. Your delineation should again follow the purpose: Which data should fall under your Metadata policies, and which data is to be covered by your policies for transactional data?

(iii) **Classification of Big Data**

Similarly to Metadata about imaged documents, unstructured Big Data files require Metadata as well, such as date, origin, or character set used.

These Metadata elements are required to accurately interpret the content of the data in subsequent steps.

In this sense, the term "unstructured" is not really adequate even for the very first raw version of a Big Data repository. It will then get more and more structured through first analysis until you have determined enough Metadata for it to become subject to Analytics algorithms, for example, for pattern recognition or correlation discovery.

Furthermore, Metadata discovery helps ensure that any content of a Big Data repository matches the data used to train a Machine Learning algorithm you want to apply.

Effective Metadata management helps do all of this in a structured way.

# Managing Masterdata
## Cross-functional MDM

Masterdata is mostly cross-functional. In other words, more than one team or department depends on a particular set of Masterdata. Yet many organizations still have different pockets of Masterdata maintained autonomously by different teams.

# The Data Model

A thorough Data Model is essential for the design of Masterdata.

Metadata is closely related to the Data Model. The cardinality and format of attributes are to be maintained in the Data Model, and Metadata management has to take this information from the Data Model.

But the structure and logic of Masterdata and Reference Data are to be defined in the Data Model as well.

As a general rule, all business logic and structure should go into the Data Model first. Masterdata and its maintenance are then defined based on the Data Model.

# History view of Masterdata

Masterdata, Reference Data, and Metadata all require a historical dimension: The changes to the value of an attribute over time.

You should always be able to replay a situation for a given date in the past (and even for a date in the future if you maintain Masterdata in advance). You may require this ability for legal reasons, for debugging purposes or to play with scenarios, to name just a few cases.

That is why web services should always offer a variant of the call where a data is provided and where the web service returns the result for that very date. This helps you test with past and future data, replay cases to gauge claims, and validate error messages.

Proper history also gives you an additional mechanism for data validation. Examples:

- You cannot have a product produced in a facility while the latter had the status "under construction." If such a combination is found, it can be flagged to a user.

- You cannot ship to a postcode that is no longer (or not yet) valid. A history view would be able to add history information when rejecting such a postcode: "No longer valid" or "Only valid as of…" provides more information than "Not valid!".

- Even a currently solvent customer may require close monitoring (or be excluded from credit payment) if that customer's credit rating has been bad for most of the past two years.

But you will also face challenges when dealing with history data:

- Year-on-year comparisons after changes: If you restructure your customer segmentation, it becomes difficult to determine the year-on-year development of customer loyalty per segment.

- While NoSQL databases can usually handle changes to the data model without a re-design, SQL databases easily run into problems. They will not recognize additional tables, miss lost tables, and are sensitive to changes to the primary key structure.

- Even where a database can handle structural changes, the client application may not be able to do so – unless it has been consciously developed to work in different modes. Unfortunately, such an ability is often hard-coded. More often, an application is modified to support a new logic, so that it loses the ability to support the previous logic.

These cases illustrate how important it is to add backward compatibility of software and databases to the list of requirements of any change – be it a big release, be it a sprint of an Agile project.

# MDM and Masterdata software
## Masterdata design styles

There is not just one way of designing and implementing Masterdata. Depending on your – functional and cross-functional – business requirements, you can choose between different Masterdata design styles for your entire Masterdata environment or for parts of it. You might decide for hybrid solutions either temporarily or as part of your long-term target.

Typical Masterdata design styles are[3]

### (i) **Centralized**

A fully centralized solution provides the safest method from a Data Quality perspective. It makes it easy to organize the Single Source of Truth principle – most of it is already part of the database design.

---

[3]Those styles and their names are not standardized. You find different names and different delineations of styles in literature. MDM solutions come with their own structure and verbiage as well.

Note that not the maintenance is centralized, only the data itself. Data Stewards from all over the world can maintain data directly in the central solution, whether it is run on-premise or in the cloud.

(ii) **Virtual centralized**

Technological progress during the past 20 years has increasingly decoupled the logical structure of a data repository from its physical design. Most database solutions allow for a distributed database setup where either the records are spread across multiple remote locations or where the database automatically replicates all records between different instances of the same database.

Cloud-based database solutions don't even know any other way of working.

As a consequence, a worldwide organizational footprint is no longer an excuse for independent, locally maintained databases. Only areas with limited connectivity may justify local data repositories – but these should ideally be read-only replicas or caches of a single logical database.

A word of caution: A primary key in a single physical or virtual database does not prevent duplicates. Typos and different ways of spelling may lead to multiple records pointing to the same "thing," undiscovered by a database's string comparison capabilities. (See also Chapter 10's section "Keep and get data clean.")

(iii) **Distributed read**

Sometimes legacy applications cannot switch to consuming data from modern MDM solutions, for example, where Mainframes are not able to call Masterdata web services or where the risk of touching an old system is considered too high.

In these cases, you would try to at least switch all maintenance of that area of Masterdata to a centralized solution (unless current Masterdata maintenance is a nonseparable part of a monolithic legacy system). You can then set up a replication mechanism to regularly update any local repositories for local consumption.

Most professional MDM solutions will be able to support all of these styles. But you usually find constraints within your legacy infrastructure. That is why an important selection criterion for MDM software is the ability to coexist and interact with legacy solutions.

### (iv) Distributed read and write

If technical constraints prevent you from centralizing Masterdata even from one single type, you could be forced to keep it in multiple repositories. In this case, you will need to ensure regular synchronization between these repositories to come as close as possible to a Single Source of Truth situation. Processes should assume, though, that data repositories are temporarily out of sync, due to the lack of real-time synchronization.

This approach requires a central registry of Masterdata sources, their interdependencies, as well as their update and usage frequency so that you can systematically automate the synchronization process.

### (v) Independent

This is the most frequently found design style in a legacy environment. It usually develops over decades in an organization with ungoverned data handling, where Masterdata has often been specified and implemented as a side aspect of isolated functional implementation projects.

The advantage of this style is the ability to implement local or functional solutions quickly – but the myriad of disadvantages makes you want to shift toward other styles.

In some cases, however, data maintained at different places are indeed logically independent. This is usually the case where specific Masterdata is required in a limited context only.

In such cases, there may not be a good business case for the integration of such Masterdata into an organization-wide Masterdata environment. You should not give up, though, but wait for an opportunity, for example, the replacement or modernization of the respective functional solution.

As long as an organization is forced to live with this Masterdata design style, the lack of technology support has to be compensated by strict governance to ensure manual avoidance of Data Quality problems.

This labor-intensive manual Data Quality work requires additional resources. You might want to leverage this fact for a business case to move to more sustainable Masterdata solutions.

# Understand your requirements first

Even if many software packages are labeled "MDM," this acronym stands for a way of managing your Masterdata, not an IT solution.

The consulting firm Gartner brought it to the point: "MDM is a complex, costly undertaking. As a technology-enabled business initiative, software alone cannot meet the challenge" (Parker & Walker, 2019).

That is why I strongly recommend that you look at selecting your Masterdata software as the very last step in your Masterdata journey. It is good to be aware of the off-the-shelf solutions' capabilities early, but these capabilities should not determine the business requirements.

If you don't know what Masterdata software is expected to do, and if you don't have the governance in place for people to use that software, you run the risk of not selecting the best possible software solution. Even worse, the solution will probably not be used properly - if at all.

# Determining your requirements

Here is my estimate based on experience across a variety of organizations: 80 percent of an organization's requirements toward a Masterdata solution can be determined up front, through the analysis and development of a Data Management framework.

Interestingly, those 80 percent of your requirements include many decisions you can take without having to choose between right or wrong (as either direction would work).

Other aspects that can (and must) be clarified up front are the number of users and applications that require access and how they will access the data. That is why data maintenance processes need to be defined early. Furthermore, you have to determine any constraints in the application landscape – and this is not limited to legacy applications.

Only the remaining 20 percent of your business requirements develop while using a Masterdata solution. And, again, most of these 20 percent are improvement ideas rather than cases of "do or die."

The simple reason is that most providers of Masterdata solutions regularly incorporate feedback from their customers. Unless your organization is entirely different from all others, it will probably not come up with any requirements that no other organization has come across before. For that reason, you can expect several MDM solutions to come without a real "show-stopper" for your organization.

In response, you should familiarize with the offerings from the leading Masterdata solution providers early in the process – without necessarily starting to build a shortlist already.

Instead, the capabilities and features of these solutions can help you shape your target setup, consisting of governance, processes, and solutions to support them.

# Waterfall or Agile?

Which methodology would you select – waterfall or Agile? Can you do your requirement gathering entirely based on Agile?

It goes almost without saying that the continuous improvement of an implemented solution should be made in an agile way by default.

Equally obviously, the process of selecting a solution needs to follow at least partially a waterfall approach, as you have to take a directional decision at one point. (You would not want to switch between MDM tools too frequently.)

At the same time, the development of your governance framework and your Masterdata processes should happen in an agile way. The documentation of requirements toward a Masterdata solution can be a welcomed side effect of teams working genuinely on Agile.

You can even develop your Masterdata governance and processes using Agile.[4] In this case, the user stories for your governance and processes can be considered user stories for your Masterdata solution as well. And your agile backlog can become part of the requirements statement for the determination of your Masterdata solution.

---

[4]You may want to ensure the product owner is a member of the Data Office to avoid functional bias.

# Build or buy?

For decades, organizations were forced to build Masterdata solutions on their own. The available solutions on the market were simply not mature enough, and they were fragmented: Multidomain Masterdata solutions did not exist, and hardly any of the solutions integrated well with each other, let alone with other parts of Data Management (e.g., with Data Modeling tools).

Meanwhile, the situation has improved significantly. Solution providers benefit from the fact that Masterdata solutions are relatively industry independent. That means providers don't need to spend too much energy into developing dedicated solutions for different industries.

But even those increasingly sophisticated Masterdata solutions come with a key disadvantage, compared to in-house solutions built to match the specific demands of the organization: If a feature is not available, you cannot simply add it – you have to hope that the solution provider considers it a good (and urgent) idea as well.

Dependency on a solution provider is a challenge, particularly when working with Cloud-based solutions where configurability is usually limited to noncore aspects.

In case you currently have one or more legacy Masterdata solutions in place, it may make sense to keep them for the time being and to introduce governance and processes to work with those legacy solutions first. You will probably learn a lot here that helps you select an excellent professional solution later.

To decide between build or buy, you should honestly assess to which extent your organization's Masterdata handling is intended to become a competitive advantage or "just" a professional activity in the background that helps your organization concentrate on their business objectives.

There is nothing wrong with the second approach! Its advantage is that you can select a Masterdata solution and develop parts of your governance processes to match that solution.

The process of selecting a solution becomes easier as well: You can determine a list of options that match your core requirements, and then you can go for the one that offers the best value for money.

# Data Governance

*"Your way sounds data-driven and rational.*
*Let's do it my way."*

**Figure 6-1.** Data without governance is a noncommittal recommendation

© Martin Treder 2020
M. Treder, *The Chief Data Officer Management Handbook*,
https://doi.org/10.1007/978-1-4842-6115-6_6

# Shape a set of Data Principles

In most cases, the direction of a data-driven organization will be different from the organization's previous direction.

Next to explaining the big picture, a Data Office team will need to judge in very concrete cases whether a proposal is in line with the data strategy, whether something needs to change, or whether something is missing.

A team of data experts would probably be able to judge properly in every single case what is right and what is wrong. And a well-documented technical Data Architecture will certainly help IT get it right.

But the general philosophy may require a generic direction that can be understood by everybody, from the nerdiest database expert to the Head of Marketing.

Jeremy Cohen from CBS used the following analogy at a data conference in London in 2018: "Compass over maps! We don't have all the answers."

A set of data principles can be your compass.

Consider such data principles the "constitution" of data. It can provide direction in new cases not yet covered by standards and policies.

But don't simply write your principles on your own and impose it on others – even if you think you know how to do it!

You might rather wish to develop these principles together with key data players across all functions in business and IT. (Please expect the right people to come from within your data network.)

It is absolutely okay to create an initial draft, on your own, or with the entire Data Office team to ensure the discussion takes the right direction. You can already incorporate points you have recognized as critical in your organization. Typical candidates are

- All data has an owner.
- Focus on most critical data first.
- Focus on root causes rather than symptoms.

But it is equally important to involve critical people as soon as possible, for example, through workshops, so that they co-own the result.

While you should not accept fundamental deviations from the data strategy, you should be open to changes that "don't hurt," such as changes to wording: "You think your alternative wording can be understood more easily? Okay, let's go for it!"

Sometimes people come up with entirely new, great ideas. If you incorporate these ideas, the result will not only be better, but you will also win those people as supporters!

Of course, you would formally own the document at all times, which allows you to fine-tune it, bringing various contributors' input in alignment with your strategic targets.

Finally, please mention all contributors in the resulting document. This move will truly make it "theirs."

Of course, you don't want to create a long, difficult-to-read monster. So, concentrate on headlines – background information can be added so that it can be read by those who want to understand better.

Search for a catchy title that you can refer to, whenever you publish a compliance check of an initiative or whenever you explain the direction. Think of "Data Tenets" or "The 10 Commandments of Data."

You find a typical list of data principles for an organization in Figure 6-2.

PRINCIPLE 1: **Focus on Business Opportunities**

PRINCIPLE 2: **Data is Cross-functional**

PRINCIPLE 3: **Single Source of Truth**

PRINCIPLE 4: **Minimize Duplication of Data**

PRINCIPLE 5: **Harmonize Data Structures**

PRINCIPLE 6: **One Common Language**

PRINCIPLE 7: **Adopt Industry Standards**

PRINCIPLE 8: **Single Foundation for Analytics**

PRINCIPLE 9: **Sustainable Data Quality**

PRINCIPLE 10: **Standardize!**

**Figure 6-2.** An example of Data Principles

# Develop data policies

## What are data policies good for?

Data policies are important to guide employees where you cannot (or do not want to) regulate each individually possible case. They are norms that regulate the handling of data, along the entire data supply chain.

Furthermore, Data Principles and policies together should enable cross-functional teams to derive concrete rules to cover known cases, in the form of processes or roles and responsibilities.

As a rule of thumb, you need a data policy wherever there are good reasons to assume that people in the organization would not deal with data in the best interest of the organization. Typical motives are laziness, ignorance, complacency, selfishness, or thoughtlessness.

A data policy intends to bring people's behavior in line with the organization's objectives and values by telling everybody what (not) to do with what data and why.

## How individual do data policies need to be?

Unlike data principles, policies need to consider the entire organization. They have to stand in the context of the overall strategy, and they must fit with the rest of the organization's policies. It does not make sense, for instance, to base your data policies on individual decision rights if all the other policies of your organization rely on cross-functional approval processes.

Secondly, the content of your data policies needs to be specific to your organization. Don't reuse another organization's data policies. The risk of them not meeting your organization's need or of you picking a policy style that doesn't match that of the other policies is too big.

Furthermore, other organizations' data policies may have been based on a different country's legal situation; they may be too specific or too general, depending on the geographical area of applicability.

However, the need for an individual set of data policies for your organization should not prevent you from checking several existing policies for aspects you may have overlooked in your draft.

## How do you determine responsibilities for policies?

As part of setting up the management of data policies, you might wish to clarify responsibilities as ownership is not always intuitively clear.

Sometimes the responsibility for one of the typical data policies is already with another team.

The most frequent example of such a policy is probably the Data Protection Policy. Other typical policies with ambiguous ownership are those that deal with data while enforcing legal requirements such as GDPR, SOX compliance (Sarbanes–Oxley Act of 2002, United States), or HIPAA.[1]

If another department traditionally feels responsible for *all* policies, including the data policies, it may be a good move to ask for the entire policy ownership topic to be clarified as part of your CDO mandate.

To do so, you would ideally create a list of all policies you would like to design on your own, plus another list of policies for which you think you should provide input.

Your main objective should definitely *not* be a maximum degree of responsibility. Instead, you should aim at the ability to establish a strong foundation for data management in your organization.

That is why it usually makes sense not to fight for the ultimate responsibility of each policy. It is often sufficient for the Data Office to be a recognized stakeholder. If users are unhappy with the original ownership, let them articulate it.

It can, therefore, be absolutely acceptable to have the responsibility for all policies with another department (e.g., Legal or Communications), while the Data Office is responsible for the content of data policies.

## How do you determine the setup of your policies?

Organizationally and structurally, you should follow the way your organization handles its policies. It must become evident that data policies do not stand outside the organization's catalogue of policies.

Furthermore, being fully in line with the rest of the policies allows you to reuse the existing infrastructure, from legal review and publication to enforcement.

Your organization may have central policy management. In this case, the data office needs to become a recognized party, owning some of the existing policies and being a formal contributor/approver to others.

If there are no policies in place in your organization, or if your organization does not have a systematic approach to dealing with policies, I encourage you to start on your own. You could even establish a program that becomes the blueprint for other teams' policies.

---

[1] US Health Insurance Portability and Accountability Act

## How do you develop a set of data policies?

As with data principles, it is essential to have all relevant stakeholders involved in the development of data policies. Those stakeholders consist of policy users (all departments) and topic owners (e.g., Risk Management or Data Privacy Management).

It is paramount to listen to the concerns of the former, and you need to avoid competing with the latter in writing policies.

Such a stakeholder-driven approach in developing policies should, of course, not prevent you from doing thorough preparatory work. You would ideally prepare a higher number of policies than finally targeted at. Teams are usually happy if they don't need to start with a blank sheet of paper, and you can somewhat set the agenda.

## What does a data policy look like?

You will not find an example here, for the following three reasons:

a) The Internet and literature are full of examples, and you should check many of them.

b) Any example I would pick could be misunderstood as the way I suggest you write your policies. The truth is: There is not a single valid (or even best) structure of a data policy.

c) Your data policies should be in line with your organization's other policies, in format and structure. Organization A's successful data policies may not work in organization B at all.

I recommend that you always have the discussion started with the purpose of the policy: What are the targeted results? Which situation should a policy lead to or prevent from happening? Which (internal or external) regulation does the policy intend to enforce?

With this in mind, you can decide whether a particular policy should be binding, conditional, or recommendation. The latter requires a reasonable degree of justification within the policy, as it aims at being adhered to voluntarily.

If you have a significant degree of freedom in shaping your data policies, you might consider using the structure of your data principles (as discussed earlier in this chapter): Each principle can come with one or more data policies that concretize it.

All policies need to be globally approved, either by the Management Board or by an adequate approval body (as discussed later in this chapter) that has been empowered by the Board to take such decisions.

Eventually, policies need to be shared, and people need to get trained. In both aspects, you would reuse whatever your organization has put in place earlier to handle policies.

Finally, please encourage feedback. You want to know whether people accept your policies.

# The target state of managed data

There are different ways of listing preconditions for well-defined data. One publicly available list can be found under GoFair (2019).

Another one is the "USA" criteria (Figure 6-3).

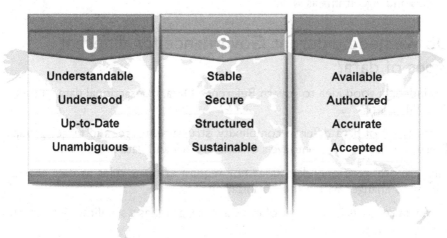

**Figure 6-3.** Data criteria (example)

While the fundamental aspects are always similar, you might want to develop your own model. This will allow you to stress those points that are of outstanding importance for your organization.

# Scope of Data Governance

Which aspects of handling data within an organization should fall under a universally valid Data Governance?

The short answer is "all aspects."

Here are a few typical questions and answers.

## Can data be too confidential to be governed?

Well, data can be too confidential to be shared. In this case, the Data Office would focus on defining how sensitive data needs to be governed, who in particular is responsible, and who gets access to that kind of data. The confidential data itself does not need to be made available to the Data Office.

## Shouldn't we exempt research data?

Freedom of research is often postulated as a precondition for creativity.

And, indeed, researchers should have all the freedom they need to discover new approaches and insight.

However, as soon as results of research activities are shared, it must be necessary to comprehend the process to get there.

That is why Governance and Documentation are important for science models and algorithms as well.

## Do we need different Governance for different types of data?

It is indeed a good idea to govern Reference Data, transactional data, and Big Data differently.

These types of data differ in complexity, structure, usage, and target group. There will also be different levels of data privacy and data protection.

Furthermore, read-only data does not require as many rules as data that is subject to changes within the organization.

Metadata requires its own set of rules as well, as it has a significant impact on interoperability.

## How about data we don't know yet?

The place to govern data aspects we are not aware of today is a policy. All data policies should be written in a way that everybody can derive the handling rules for any new data aspects from these policies.

If, for instance, a new law asks for a change within the next two years, a policy could state that any new law needs to be complied with at least two months before it takes into effect.

And if there is no policy in place for a new topic? Use your Data Governance bodies and processes to address any new issue, as sustainably as you can! And for this approach to be followed, make it mandatory – yes, through a policy!

---

## DATA MANAGEMENT THEOREM #9

Data Management requires centralized governance.

Execution should be delegated by default.

---

# Decision-making and collaboration

What could work better than casting all responsibilities into a well-defined governance model, with cross-functional representation across all hierarchy levels?

A good starting point can be a high-level model, as shown in Figure 6-4.

Review and Decision Process

4. **Management Board** as final **escalation** body, as well as for **strategic Board-level** decisions

3. **Executive Data Decision body**
   2nd hierarchy level; cross-functional!

2. **Data Collaboration group** on 3rd hierarchy level, to develop concepts, review, propose, decide

1. **Data Community**: Subject matter experts work together, exchange knowledge, make proposals

**Figure 6-4.** The review and decision ladder

If you begin with four different levels, you can subsequently define each of them in detail, adjusting it to your organization's structure and need.

# Management Board

No organization should keep the Board out of its data governance structure.

I have seen many cases, particularly in big corporations, where the Board was considered too strategic to deal with "ordinary" data topics. This is a missed opportunity!

Whether a Board consciously decides about data topics or not, their decisions are intertwined with the organization's data. They decide about data, whether directly or indirectly, whether consciously or unconsciously.

Prof. Dr Jacques de Swart, Partner Data Analytics at PwC and Professor at Nyenrode Business Universiteit, put it straight, during a data conference in Amsterdam in 2019: "The big decisions are being made in Boardrooms." You need to get your big data topics there.

We need to be clear here: The Board should not be expected to discuss regular data work. However, it needs to have two critical roles in an effective data governance model:

### (i) Strategy

A data strategy cannot exist independently of the overall strategy of an organization.

While the responsibility for the development of a data strategy is with the CDO, its endorsement through the Board is indispensable. The same applies to the Data Vision and the Data Mission.

Without a clear statement from the Board that the data strategy is fully supported and considered in line with the overall strategy, the CDO will lack authority in execution.

Whatever the CDO does, it should be possible to relate it back to Vision, Mission, and Strategy: "We are doing A to achieve X as approved by the Board."

Hand in hand with approving the data strategy, the Board should be kept up to date regarding the execution progress. In other words, regular brief updates to the Board need to be part of the overall Data Governance. This keeps Board members aware of the subject, and it opens a channel for the CDO to present ideas or express concerns.

### (ii) Escalation

Following the subsidiarity principle described earlier, each hierarchical level should resolve as many disputes as possible before escalating up.

Furthermore, disagreement between two parties should be resolved bilaterally instead of involving all functions. The same applies to the Board where the CDO can moderate an agreement between two disagreeing Board members.

However, each organization needs a formal process to resolve the remaining topics as well. Experience shows that minor issues don'tmake it to the Board agenda even if there is no agreement on the levels below. There should be multiple options to resolve such issues at any level.

# Executive data decision body

Below the Board, you will usually need an authoritative body to decide cross-functionally in data matters. Only small organizations would ask the Board to take such decisions.

For such a body to have sufficient authority, it needs to represent all areas of the organization. This does not mean that the entire second management level needs to be on this body. Instead, selected, data-savvy executives may represent multiple areas – following previous alignment and authorization by the Board.

It is often a feasible approach to have each Board member select one direct report to represent that Board member's entire area of responsibility. This can at least be a starting point, allowing for further fine-tuning, for example, where two subfunctions under one Board member are too independent for one executive to represent both of them.

I am consciously using the term "decision body" here as I don't expect this body to *discuss* topics in detail – unless they really want to. In all other cases, I'd expect the teams involved in data-related discussions to agree on proposals up front and to prepare their body members about any agreement as well as all remaining points of disagreement.

As part of the necessary data escalation process, authority for any escalation to the Management Board should be with this body only.

I call this body the **Data Executive Council**. You may wish to follow the naming conventions of your organization if any.

# Data collaboration group

For the Data Executive Council to be able to decide, you need a body that prepares those proposals, requests, or escalations content-wise.

This should be done by a cross-functional group of team leads or department heads.

I'd call this body the **Data Management Council**, or shorter, the **Data Council**. It should be chaired by the CDO.

This council should meet regularly to devise concepts, review, propose, decide, or escalate:

- The Data Management Council reviews determined Data Quality issues.

- It decides about new topics to be addressed by the Data Office or to be worked on by cross-functional project teams.

- It asks for updates from the Data Office about their plans, activities, and progress. The CDO can ask own team leads to present to this body.

- It reviews, preprioritizes, and decides about project proposals into the organization's process for prioritization or budgeting of projects.

- Finally, any requests for the Data Executive Council go through this body.

Members should be appointed by the members of the Data Executive Council. Data Champions are usually a wise choice.

While the Data Executive Council benefits from a limited number of members, the Data Management Council might benefit from being bigger in size, to ensure a truly cross-functional discussion and coverage.

## Data Community

Data experts are ideally spread across the organization. Each department should "speak data." See Chapter 9 for different possible roles.

In addition to them being part of their respective teams, they should form a virtual community that, among others, serves as subject matter experts for the Data Management Council.

People don't need to be appointed members of the Data Community; everybody whose role includes a task to maintain, protect, share, improve, or interpret data assets should be considered part of the Data Community.

However, this community requires active moderation so that its members can work with each other and learn from each other.

The community should allow those data experts to bring in new ideas and support data projects across business functions. The moderator of this community should be a member of the Data Office who also serves as the bridge into the Data Management Board.

## Data review and decision process

Before detailing out all data processes (which we discuss in Chapter 8), we need to deal with the mother of all data processes, the "Data Review and Decision Process."

This process describes how topics go up and down the review and decision ladder between the subject matter experts and the Board. It also covers the handling of all other data-related processes.

# Speed of implementation

Data Governance cannot be implemented overnight. This is less a matter of complexity than a challenge to get everybody's buy-in.

Even if you know where you are heading, it is advisable to break your journey down into small steps. Get people used to see data governed.

You can also expect to learn as you take the steps. What you would have expected to work well may turn out to be impossible. Alternatives may come up and change your original ideas.

Or, as Laki Ahmed, VP Global Head of Enterprise Information Management at Signify, said at a data conference in 2019, "Allow your governance framework to evolve."

As soon as you have the four levels of your data governance ladder defined, the key parts of it implemented and supported by a mandated review and decision process, you have a basic Data Governance structure in place that allows you to work on all remaining topics.

# The Data Language

*"The meeting is running late, so I propose that we debate what we mean by 'dashboard' and put off the discussion of our critical information needs to another time..."*

**Figure 7-1.** "Do we need one language?" – "Define 'language'..."

© Martin Treder 2020
M. Treder, *The Chief Data Officer Management Handbook*,
https://doi.org/10.1007/978-1-4842-6115-6_7

# Characteristics of language

## Don't we all speak English?

For people to understand each other, it is not enough that they sit in the same room and that all of them can speak and hear.

This statement becomes obvious as soon as you think of someone from Greece and someone from Indonesia trying to communicate with each other.

The same applies to different industries, organizations, and departments. And just as an Indonesian person can learn Greek and use a dictionary to look up words, we need language training, a glossary of business terms, and precise definitions on how to deal with this language.

Such an overall setup consists of the definition of expressions (the data glossary), Data Rules, Data Standards, and the Corporate Data Model. This chapter describes an overall framework to define an organization's data language.

## The dynamics of language

Language is not static, as it needs to reflect reality. New things come up that require names. New activities call for adequate verbs.

In response, people reuse existing words and descriptions from other areas of life, or they borrow words from other languages. Sometimes they even artificially create new words, often based on ancient languages or abbreviations.[1]

Here are three typical examples around the English language:

---

**EXAMPLE 1**

---

There have been chains and there have been saws for centuries so that there was no need for a new word when the "chain saw" was invented.

---

[1] This approach, called neologism, is generally a normal process in the development of any language.

## EXAMPLE 2

In agriculture, people had used "tractors" (Lat. *trahere* – to pull) well before the industrial revolution, so that this word was applied when the first engine-powered devices dug over our fields.

## EXAMPLE 3

The ancient Romans weren't particularly good in maths. You know that if you try to multiply two numbers written in Roman numerals.

But they had to calculate and to add figures. The verb they used to describe this activity is "computare."

When the first machines came up that took over this work from humans, it seemed natural to call them computers.

Whenever an entire ancient language such as Latin or Hebrew had to be prepared for modern usage, the process of enhancing its vocabulary had to be executed systematically to fill all gaps consistently. Without such a coordinated approach, it sometimes takes centuries until newly created words are documented, standardized, and fully adopted.

Now, what does this history of language have to do with data?

The habit of reusing existing terminology for new situations or of creating new expressions based on existing words is not limited to our private world. It is frequently applied in our everyday business life, resulting in a risk of ambiguity, across all business areas.

And all of this applies to the world of data as well. Furthermore, this data world is still relatively young. There has hardly been any time for a data language to develop, resulting in a lack of historically developed expressions for many situations and circumstances.

We obviously need to do something to address these issues of ambiguity, misnaming, and lack of clear designation. This applies to the language of business in general and to the language of data in particular.

# The data glossary

## What is a glossary?

A glossary is a repository of terms, together with their definition or explanation.

Modern glossaries come with cross-references, synonyms, history, hierarchies, relations, and ownership.

Technically, good glossary tools have a web front end, a search function, and a change workflow including proposal functionality.

## The risk of not having a glossary

What are your organization's ten biggest customers?

Easy question?

Do you have the answers to all of the following questions?

- What is a customer? Is it a "company" (like "Microsoft")? Is it an account (everything covered by the same rate card)? Is it the logistics procurement person our salespeople talk to?

- Are a customer and an account the same thing?

- How many different definitions of account do we use?

- Do we add all organizations to one customer that belong together? How do we know?

- Do we consider Volvo cars and Volvo trucks part of the same company?

- What does "belong together" mean? 100 percent? Majority? Joint venture?

- Which shipping period do we choose?

- Do we rely on data from external providers like Dun&Bradstreet? Do we know *their* definition?

Would you agree that a seemingly straightforward question like the one for your biggest customers does not necessarily have a clear answer?

But does this mean that we have to give up, as there is no chance to get it right? Are there too many options?

Well, it is a matter of asking the right question!

Ask yourself: What do we want to do with the information "Ten biggest customers"? This question may help define both "customer" and "big" – in this specific sense.

---

## EXAMPLE 4

"One customer" can be

- All shipping volume that would be at risk if we messed up in a certain case. Caveat: In some cases, even the regional organizations within one organization don't talk to each other in the area of logistics.

- All entities under the same CEO: This is important if we want to build a personal, board-level relationship. But how would we deal with 50 percent owned organizations? And shouldn't we also have a look at the entities (investment funds, pension funds, etc.) that hold shares of most of our customers?

- All business under the same buying decision: How do we determine on which level these decisions are made? Maybe an organization's CEO doesn't care about smaller purchase decisions, leaving these decisions entirely to the individual subsidiaries or divisions.

---

Each industry has its specific language. The different organizations within an industry have developed their specific verbiage. Different departments use different terminology. Employees give business expression a different meaning based on their personal background. I am sure you could add to this list, based on your own experience.

All of this leads to the following two results:

- An object is called differently by different people.
- The same expression is used for different objects.

While the former is just annoying (you might have to ask to understand), the latter can be tremendously dangerous as people may discuss for a considerable time without realizing that they are talking about different objects.

I recently talked to a CFO of a larger freight forwarding company. He didn't see the need for a centrally governed glossary. I asked him whether, in his company, Finance and Sales have identical definitions of "revenue." He replied that this would never be allowed and that Finance is, of course, the only department that can define "revenue."

While I congratulated him for his accepted business ownership of this term, I asked whether the Finance definition of "revenue" was documented and shared with the rest of the organization. He assumed the former but didn't know the latter.

I explained that the Sales department would require an expression for the monetary equivalent of their monthly sales. I said I suspect they use the term revenue as well, just without Finance being aware. Yet they would probably define it differently: In order to calculate the bonus of their sales reps, they'd add up the value of all sales deals closed in a particular month.

The CFO said "No, that's wrong! It's when the service is rendered that we count it as revenue!" I said "That's the point! As soon as there may be different parties using a term differently, there is a good reason for a neutral body to orchestrate."

The CFO may indeed be the owner of "revenue" – but in this capacity, his department would need to seek the dialogue with all other departments to ensure all variations are properly defined and named unambiguously.

In an organization with proper data management, all of these different definitions will get discussed, developed and documented as part of an organization-wide glossary.

A glossary is also a precondition for unambiguously written processes and policies.

Furthermore, the relatively young "data language" comes with expressions that have already been in use in other contexts, lacking existing verbiage to cover them.

As a result, business documents can easily be misunderstood as the words and expressions used here could have multiple different meanings. This is getting worse with an increase in data-related content.

What are typical cases of confusion?

It is already problematic if someone uses an expression that this person's dialogue partners don't understand. As mentioned earlier, it is even worse where two parties use the same expression for different purposes. I have witnessed meetings where people talked to each other for hours before they (and I) realized that they were referring to different things. Sometimes they even found out long after the meeting, with significant impact.

To make this situation more tangible, let's look at a few business examples from the Human Resources area.

---

## EXAMPLE 5

As part of a reorganization, HR writes about "impacted" employees. What do they mean? Some employees will interpret it as "something changes for them." Others will read "those employees get fired."

---

## EXAMPLE 6

"Headcount" needs to be defined to be mapped properly. Do part-timers count as full headcounts? Are apprentices included? How about "inactive" employees, for example, those with a long-term illness or those on parental leave?

---

Imagine three acquired organizations that continue to work with their respective legacy HR systems, and you put a joint reporting tool on top. You have to bring the hererogeneous data from these three organizations together.

Building unions from different database tables is easy, technically. But do the same labels stand for the same definitions of columns? As long as the data types are compatible, you would not even receive a warning if the definitions differ.

"Employee" is a typical entity you could define in a lot of different ways. Remember the last example above: Maybe one organization works with "full-time equivalents (FTE)," while another one works with employed persons. One organization only counts full-time employees, while the other two include part-time employees. Are contractors included? How about the long-term ill ones?

A glossary is a great support in your task to organize the data. Mapping may still be necessary, but now you know how to do it.

On top of this, many data-related terms require both explanation and definition. Look at the expression "data model," for instance, which is not the digital equivalent of a fashion model. Such a data model may describe the "relationship" between two subjects – which does not say whether they are friends or brothers.

## What needs to go into a glossary?

A glossary does not need to help determine the "right" definition. Instead, it needs to help an organization agree on an unambiguous language.

Any organization can autonomously define its verbiage – although it usually makes sense to look beyond your own organization to avoid ambiguity in the dialogue with customers, suppliers, and other external parties.

What do you need to include in a proper glossary?

- Definition
- Synonyms
- Related terms and differences
- Reference to the logical data model and to business processes
- Lineage
- Workflow for change requests

## How do you introduce a glossary?

Do you have to convince people that a glossary makes sense? Probably not. It is a more frequent challenge to unite multiple independent glossaries into one. Every department head, every project manager, and every business architect have faced problems with ambiguity and misunderstandings.

A frequent conclusion is the creation of a glossary within the remit of the respective role. You find the "Marketing Glossary" on the Intranet site of the Marketing department, you find a glossary in the appendix of most project documents, and most specialist documents such as policies or expert reports come with their own, independent glossaries.

So, how do you merge all of these into one glossary? And how do you ensure nobody creates another individual glossary in future?

It starts with the mandate: You need a cross-functional, board-level decision that glossary ownership is with the Data Office.

And, yes, this topic is important enough for a Board decision.

## Data Rules and Standards
## The purpose of rules and standards

Why am I dealing with Data Rules and Standards under "Language"?

While the line between "coining an expression" and "defining an expression" is blurred, it all starts with a common language.

In most cases, you will implicitly establish a rule to close an ambiguity gap in your organization's data language.

If you describe data entities and data-related expressions in detail, and if you extend the relationship between data entities to algorithms and dependencies, you are amid the definition of data rules and standards.

There is no academically sound delineation between rules and standards in the area of data. I use "rules" for rather technical instructions (including metadata), as opposed to "standards" which stand for regulations, mostly in the form of descriptive text. Feel free to deviate from this distinction in your daily practice.

Wherever rules and standards are not adhered to, you would distinguish between accepted deviations (Data Concessions), temporarily accepted deviations, and violations. Detailing this out is part of managing data rules and standards as well.

---

### EXAMPLE 7

A classic Data Rule would state that the format of postcodes in any address database needs to adhere to the formats published by the Universal Postal Union.[2] It would specify the level of detail (e.g., "There can be more than one postcode format per country") and how the underlying metadata is coded (e.g., as regular expressions[3]).

The corresponding Data Standard would state how frequently the metadata needs to get updated based on changes published by UPU.

Ideally, the Data Standard would even refer to the library and object a software developer is supposed to use to validate postcodes. It may state that the use of that library is mandatory and that direct usage of the metadata is not permitted.

If you have a Data Principle that requests the usage of APIs and web services instead of using raw data for the development of software solutions, you can directly derive this Data Standard from it.

---

## Data Standards

A Standard describes how something should be (or not be), as part of an agreed target of an organization's data handling.

It needs to be written in a way that you can determine compliance in an objective way. In other words, it must not be possible to disagree on whether a data standard is adhered to or not.

By the way, the first standard should describe the nature of a standard itself. The two paragraphs preceding this one already shape such a typical data standard.

---

[2]Postcode formats and further country-specific address format information are consolidated and provided by the UPU. See (UPU, 2019) for details.

[3]For a good comprehensive description of RegEx including many examples, see www.regular-expressions.info (Goyvaerts, 2019).

## Data Rules

A Data Rule describes data to the lowest level of detail, down to metadata structure.

At the same time, a Data Rule is by no means restricted to a single entity or attribute. It may, for instance, state that only two of three attributes of an entity may have a nonzero value.

But where do these Data Rules come from?

To a huge extent, data rules are based on business rules. They are translations of how an organization's business is supposed to work, in a way that allows for direct implementation in software.

In reverse, the uncertainty of a software developer on how to implement is a good indicator of a missing or incomplete data rule. The Data Office should systematically use such situations as opportunities to clarify the data rules with the business owner (forcing the latter to think about the underlying business rule).

In short, you can consider Data Rules the main input for a Corporate Data Model. A significant part of matching an organization's business view with its technology view happens when translating Data Rules into a Data Model.

## Working on Data Rules and Standards

Your data rules and standards are *complete* if they contain all information required to create all necessary data processes and IT solutions.

But please don't consider this as a suggestion to complete all of your data rules and standards before you are ready to define processes. Instead, this should be an iterative process:

- The development of processes will discover gaps in data standards.
- Closing those will allow for better processes.

You can (and should) work in different subject areas in parallel, at different levels of maturity.

And please start using Data Rules and Standards before they are perfect. In many cases, you can only determine gaps or inaccuracies by actually applying those Data Rules and Standards.

# Documenting Data Rules and Standards

Whatever the Data Office develops needs to be documented, of course. But where and how would you document Data Rules and Standards?

Unlike policies, Data Rules and Standards cannot easily be merged into existing nondata-related repositories of the same kind. Ideally, there are no other repositories of this type as Data Rules and Standards should cover the entire business.

This situation gives you the freedom to develop the required repository within the Data Office.

Data Rules and Standards are closely related to the glossary, and they are required in the context of processes. Access should be provided through the same user interface, for example, the same Data Office Intranet site.

Everything else depends on the software solution you use to document and publish data-related information. Well-integrated solutions allow you to maintain the glossary together with Data Rules and Standards.

It is important to document not only the standards and rules but also their lineage and their implementation status.

Finally, I personally recommend that no definition ever comes without examples.

# The Data Model

## The value of a Data Model

Suppliers usually tell us that strict data models are overrated and that Big Data even focuses on unstructured data. But is this true?

If we say that data is a language, then the data model can be considered its grammar.

Whoever had the pleasure of learning Latin may recall situations where knowing each word of a sentence was not sufficient to understand the meaning of the sentence. I bet this was when you understood that grammar is important!

It is the same thing with data. Pieces of data cannot be turned into information if you don't know how they are interconnected and what their structure and meaning are.

As data without structure and logic is without any value, you need a strong, well-defined "grammar" for your organization's data.

# The value of ONE Data Model

The negative impact of using multiple data models in parallel is not just academic theory. Most organizations can observe this today. History has often led to different explicit or implicit data models used by different teams.

Not surprisingly, most inconsistencies can be observed at the borders between different functions. Most functions know their own part of the data model, but they define other parts as they assume them to be.

All data consumers suffer from inconsistencies caused by the absence of a single data model:

- In Analytics, working with different data models leads to inconsistent results, even if different teams use the same data set.

- You cannot optimally run daily operations if you steer it using parameters from sources based on different data models.

- Standard reports show inconsistent messages, and KPIs are not reliable.

- Changes of the business model are complicated if you cannot base them on a standard data model.

- With data increasingly moving into the cloud and taken care of by third-party SaaS providers, data will have to conform with external data models.

- If you move toward a single source of truth, some implementations may fall over, as they are based on a different data model. Typical cases are primary key violations or duplicates. It may take months until the first "duplicate" kills the database table, but it will certainly happen over the weekend...

That is why each organization should agree to work toward one single data model. It is frequently called the **Corporate Data Model**, abbreviated **CDM**.

## Example customer data

External providers offer to manage your customer base for you. And it's easy, right? "A customer is a customer is a customer," and depending on the industry, a customer has a certain number of attributes. No open questions?

Let us validate this assumption from two different perspectives:

(i) **Customer hierarchies**

- Do you want to consider a particular organization the same customer as another conglomerate that owns it?

- How about a company owning a certain share of another company: Does it count as the same company? Maybe not if it is a small percentage only? Is there a natural threshold?

- If two different organizations belong to the same conglomerate, should they be linked?

- Can your data model handle changes to customer hierarchies (e.g., mergers and acquisitions between your customers)?

To tidy up, we need to do two things:

Firstly, we need to understand the purpose (or multiple purposes) behind customer classification: What do we want to do with the customer records?

For example: If we want to avoid inconsistent pricing toward the same organization ("My colleague from subsidiary X told me they get a bigger discount than us in subsidiary Y!"), we may have to consider legal and commercial ties between organizations.

Secondly, you'd have to determine all customer attributes that you need to classify and categorize as per your business needs. "Member of purchasing association X" could be an important attribute missing in your (or your cloud provider's) data model for "customer."

(ii) **Customer status**

Most Sales solutions distinguish between suspects, prospects, active customers, passive customers, and retired customers.

- If so, isn't any retired customer a suspect?

- Can a retired customer become a prospect again?

- How long after its last business transaction with you would you consider a retired customer a prospect again?

- Can different customers have different statuses if they are closely associated (see under (i) above)?

Now imagine you share customer data between different departments. If you don't agree on a single CDM, one department may filter out all retired customers as it focuses on Sales activities. The other departments would need them as it deals with Customs matters where customer data needs to be kept for a long time, even after the customer's last transaction with us.

Will an incorrectly filtered data flow result in visible errors? No, all processes will work seamlessly. Is this sufficient? No, it is tremendously dangerous, as it hides the data issues: Customer records may be missing, but the systems are not able to determine the gap.

## Analytics and Data Modeling

When your data modelers shape the CDM of your organization, they need to consider all base elements and all of their attributes and relationships.

On the other side, Analytics needs the freedom to create new data entities, attributes, and relationships to reveal valuable insight.

But where to draw the line?

As a guiding principle, any data logic derived from an organization's business model should be part of a centrally governed CDM.

Furthermore, Analytics folks should never redo that CDM in their visualization tools – although more and more tools offer this capability. Otherwise, an organization runs the risk of the same data being maintained (and modeled) independently in different data models.

For the same reasons, external data should never be imported straight into data visualization tools (unless it is really required for one single task only). Instead, external data needs to become part of your organization's Single Source of Truth, so that all users take the data directly from there, in an agreed format and structure.

To be successful here, you should make it convenient for data analysts to leave the data modeling to a central team – otherwise, they will be tempted to model data in their tools, in order to be in control, or to obtain results more quickly.

First of all, you need to define unambiguously what falls under the CDM: Where filtering is specific to a concrete analytics task, you would leave it flexible (e.g., filtering transactional data by a combination of attributes to determine correlations).

Whenever your analytics folks state that they need to touch the CDM to proceed, it is often the right choice to address the data sources instead. For instance, any "unknown" or "n/a" values may have to be removed from a column that is considered purely numeric from a business perspective. The result becomes part of the logical **Single Source of Truth**.

Why would you do this? Firstly, different teams doing the same filtering independently of each other means additional effort; secondly, the groups may filter slightly differently, resulting in noncomparable results.

Process-wise, any request for an extension of the core data model falls under the process group "Data Logic Changes" (see Chapter 8). Such requests are good examples of why this process needs to be very agile – data analysts don't have time for a lengthy approval process.

At the same time, a data model extension usually comes without any backward compatibility issue, so that the impact on existing solutions and databases is expected to be limited.

## Conclusions

Each organization should have one single Corporate Data Model (CDM). It needs to be tied to the glossary as well as to the business processes, and it should be binding for everybody in the organization.

While we need the flexibility to work with semi- and unstructured data or to play around with future options, our existing business model must be described by a well-defined CDM.

Such a CDM is not only necessary to avoid crashes when two worlds need to connect. It also helps prevent data losses and misinterpretation, which often goes unnoticed.

The concept of a "Single Source of Truth" applies to the format and relationship between entities as well.

To get there, you may want to apply the Data Management 101 as described throughout this book. As none of the single steps is easy to execute, having a framework helps a lot:

1) **Discuss** the challenge with all relevant stakeholders. Explain the problem and get their buy-in to the next steps.

2) **Set up** a **project** at least for the first two phases: Developing a target and determining the gaps. Get this project funded and staffed (you need resources from business functions and IT to support this phase).

3) **Develop** the ideal CDM, that is, the **data model** that accurately describes the current business situation and comes with sufficient flexibility to cover future changes. This is an activity in close collaboration with business functions.

4) **Obtain** cross-functional **approval** for this ideal CDM.

5) **Make** an **inventory** of current data models – both implicit and explicit models. Sometimes you will need to have existing applications reverse-engineered.

6) **Document** all **deviations**.

7) **Evaluate** the (current and future) negative **impact** of existing deviations from the ideal CDM. Ask the stakeholders to quantify this impact. The Data Office teams should just support.

8) Based on impact and priorities, **devise** a **road map**. It may be a multiyear road map, but it should cover the entire journey toward a fully implemented CDM.

9) Get your **road map approved**, using the outcome of step 7) as your business case. You should be able to refer to that approval when taking any of the next steps.

10) **Run** the **implementation** as single projects under one umbrella program. While the Data Office does not need to run each of the implementation projects (they might even be subprojects of other projects), you need to own the overall initiative.

The biggest challenge is to convince business folks that your organization needs to work toward a single CDM. It is paramount to help them see the value for money, the "Return on Data".

You might find this too big a program, considering the bandwidth of your organization in general and your Data Office team in particular. Alas, feel free to start small!

It is better for your team to begin with a subgroup ("customer" is usually a good one) than not to start at all. And any visible success with the first subgroup may increase business buy-in to your overall objective.

# Choosing a software solution

## Do you need a tool to manage data language?

A good tool makes your life easier. But it is not a precondition. Don't wait until you have a tool. A glossary is too important to postpone its introduction.

A tool can automate, but all activities can be done manually as well. If you start with a spreadsheet, its real-life usage may even provide better insight into business requirements on a tool than if you have to guess before you start.

The biggest advantage of a tool is the immediate online availability of all information. People tend to download files and to use them offline – this is a significant risk that comes with manual management, be it for your glossary, rules and standards, be it for the data model.

## Are there any primary requirements?

Data experts usually have a good understanding of the required functionality of a tool that manages your data language. However, they often underestimate two key requirements:

(i) **User experience**

The target group is business folks, not IT admins or database designers.

(ii) **Collaboration capabilities**

None of the aspects of a well-managed data language can be created in an ivory tower. All of it grows with everybody's input.

All in all, a tool is good if it has all the necessary technical features AND if the user experience looks like this example.

---

## EXAMPLE 8

I open my browser and click the bookmark for "glossary." I type either an expression or a description. If I find what I need – great! If not, I describe what I need.

I can work with business use cases: Instead of having to check a box, I can write "Two employees have talked past each other for one hour as they didn't realise they were using the same expression for different things" or "It is unclear to me whether the revenue calculation in context X should include internal revenue." I know that a competent human being is going to read this.

The system creates a case, and I will get a personal response within an agreed time (based on a documented service-level agreement). The case management functionality starts a process to solve the topic with all relevant stakeholders.

The system will keep me posted on any progress, and it will ensure that my case does not get forgotten.

The result including any possible changes to our organization's language (i.e., glossary, rules, standards, or CDM) will be communicated broadly, and I can look up the history of the discussions online.

You can consider this an example for use cases you can develop together with your stakeholders. Such use cases help to ensure you solve **their** business problems instead of any academic data problems.

Later, you can use these use cases to check whether your data language setup works as expected.

# Data Processes

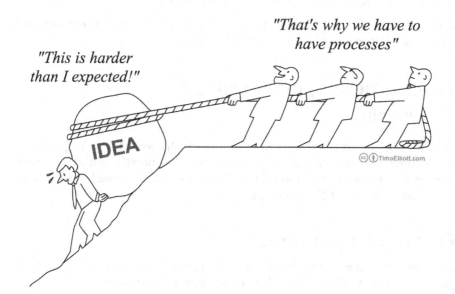

**Figure 8-1.** Processes are important

© Martin Treder 2020
M. Treder, *The Chief Data Officer Management Handbook*,
https://doi.org/10.1007/978-1-4842-6115-6_8

# Why prescribing processes?

Whether people like it or not, you need to prescribe how they are supposed to deal with data. Otherwise, you will not be able to guarantee that all tasks are addressed or that all necessary activities will be executed properly.

Processes are complex to describe – as you will not want to keep a "next step" undefined or ambiguous. However, it is definitely an exercise that pays off soon.

Given the number of processes and their complexity, nobody outside the Data Office should be expected to know all data processes by heart. They should, however, know whom to ask or where to look. See Chapter 11 for an organizational concept.

A good start for the development of data processes is your set of data policies (which is another reason to start with these early).

Each process will need to serve a purpose, that is, substantiate a policy, with a focus on delivering concrete, required results.

# Process development aspects

## Ownership

Processes are about ownership: Each step needs an owner. For an organization to become independent of single individuals, ownership should be with functions rather than with individuals. But you will always need to be able to determine the name behind a responsibility.

## Existing data processes

Don't reinvent the wheel: Where well-functioning processes are already in place, incorporate them. You will increase their acceptance.

## Collaboration

Don't develop processes behind closed doors. Start with the objectives and obtain agreement on those. This will make it more difficult for people to oppose processes that meet those goals.

# General considerations
## Technical debt handling

Tight business timelines may require workarounds to be accepted. Such acceptance, however, has to be on a temporary basis.

The resulting "technical debt" needs to be documented to ensure closure in the end. (See concession handling in section "Project Data Review process.")

## Conflict management

Lack of bandwidth, funding, experts, time, and others may result in conflicting requests. These need to be resolved by taking the perspective of the overall organization ("What is best for the shareholder?").

The "data review and decision process" as introduced in chapter 6, including its bodies Data Executive Council and Data Management Council will provide the necessary cross-functional mechanism.

You will face challenges even if the model has been set up well, but you will need to manage those challenges actively. This is why such a master process should include a generic escalation process (see also "A decision and escalation process" in chapter 2). Challenges include (but are not limited to) the alignment between entities, legal topics, InfoSec constraints, and budget and resource constraints.

## Ease of use

A key focus should be on "avoidance of bureaucracy." It must be desirable and attractive to follow this process. Otherwise, people will simply ignore it or choose informal ways of dealing with data.

That is why, for instance, users should be offered a template that "Helps them not to forget anything" rather than something that looks like additional work. And please don't make the provision of all details mandatory in the first go – you will need to talk to the requestor anyway.

## Process triggers

Each data process requires one or more well-defined "process triggers." The purposes are

- To make clear in which cases a data process gets activated

- To ensure the activation of a data process from within other processes wherever necessary
- To inform people how processes can be triggered and who is empowered to do so

A process trigger always comes with the description of preconditions and information to be provided. It must be easy to judge whether a process trigger is valid and complete.

This description needs to take a user perspective: We need to make it easy for everybody to find the right category to apply the right process.

Here's a list of typical process triggers:

## Request for project approval or funding

Projects need to comply with data requirements to get approved, be it as an initial approval or be it as a final approval of a project's deliverables.

If your organization has an ARB (Architecture Review Board), its approval should be part of the process as well.

## Issue report

If somebody reports something to be wrong, the right process needs to be triggered.

This mechanism is not only valid for business users or IT experts. It also includes any findings of the Data Office's internal Data Quality assessments.

## Change request

Typical examples of change requests pertain to data structures, terminology, applications, a data-handling process, data sources, or data flow.

## Expiry of technical debt

Each temporarily approved technical debt comes with an expiry date. Reaching such date without the technical debt being closed must trigger a data process so that the debt is acted upon. It is recommended to also have a pre-alert trigger, to notify of an approaching expiry well in time for corrective action.

## Request for clarification

Existing policies or guidelines may be ambiguous. There must be a formal process for a user to ask for clarification. This process can be very lean as it usually does not require any change to the subject itself.

## Escalation

Escalation often has a bad connotation. But if applied properly, escalation simply helps solve conflicts by elevating the decision to a higher level.

It is important to realize that, in most cases, disagreement has nothing to do with personal animosities or lack of understanding. Instead, different people have different priorities based on their respective roles.

Solving the conflict on a higher hierarchical level means taking a broader perspective: How would an owner of our organization have decided?

To keep escalation free from emotions, a well-described escalation process helps a lot.

# The data review gate

Criteria for activities to have to pass through the data review gate:

(i) **Potential changes to the structure or logic of data**

For example: Data field mapping between different departments; changes to the structure of the attributes of "customer"

(ii) **Introduction of changes to data handling (processes, roles)**

For example: Masterdata management as part of the introduction of a new MDM solution

(iii) **Potential changes to the data life cycle (sourcing, movement, modification, consumption)**

For example: Replication of Masterdata into the Cloud (Oracle, Salesforce)

(iv) **Suspected cases of violated Data Standards or Data Principles**

For example: Duplicated Reference Data maintenance, data sourcing from an unreliable source

This set of criteria should allow us to keep the process leaner, by always asking only for the information required for the specific category or entry point.

# Concrete process groups .
## Data request process

A data request process describes how someone can obtain data or access to data in order to work with it. This covers every type of usage, from support of operational processes to data science, be it the request for a single record or a set of transactions, a one-off request or permanent data access. This is not unnecessary bureaucracy, as uncoordinated usage of data by business functions is the number one source of data misinterpretation.

Such data requests from all functional areas need to be coordinated and aligned in a very transparent way, with short response times. You don't want projects to experience delays waiting for data...

If necessary, requests need to be prioritized. The process should, therefore, define criteria based on effort and benefit.

For the Data Office to be able to coordinate this activity, it needs to establish formal interaction with all business departments to determine their concrete data requirements.

These requirements include

- The requested kind of provision (access, format, etc.; one-off versus repetitive; full versus incremental)
- Timeline and urgency
- The content of the data
- Dependencies
- Required quality

Evaluating these requirements will lead to a set of data-related activities, responsibilities, and milestones.

If direct sourcing of data turns out to be the best approach, it may be decided for – but transparently, with all risks being managed.

All data requestors are expected to always talk to the Data Office first. They need to share what their business objectives are and what their intended business processes look like.

In response, the Data Office needs to share targets, guidelines, and governance wherever these are already available. Workstreams are then expected to perform gap analyses.

Certain deliverables (solutions, data) may already be available. They need to be made visible in a way that data requestors can check on their own – which will allow focusing on the true gaps.

**Don't overdeliver!** Sometimes a data requestor does not require high-quality data. In such a case, a data provider should not insist on higher quality.

At the same time, the Data Office also has an advisory role. Data requestors may underestimate the long-term impact or side effects of accepting low quality. Any such request requires a conversation with the data experts to allow for trade-offs before proceeding.

A proper setup will allow you to react properly to each request for data or data-related services:

- Data is already available? Take it from there.

- Data can be provided by one of the data provider teams within the given timeframe? Do it!

- Data activities cannot be parallelized for technical reasons? Prioritize.

- Data provision requires more funding or resources? Prepare a funding request.

It will also allow you to distinguish between

- Short-term, project-related data with low accuracy needs (e.g., supporting planning)

- Long-term, final, project-agnostic data acquisition (shaping the long-term organization)

You need to be able to increase Data Quality:

- Data Modeling and data mapping need to be applied consistently across all functional activities.

- All data sources need to be agreed (Single Source of Truth).

- Short-term gaps need to be made visible and covered in the long-term planning.

## Project data review process

Most projects require and use data, and you need to ensure they do it the right way. That is why you need to define data control points throughout a project, starting already with project preparation, to catch noncompliance early and to give the requestors the chance to redo their proposal, ideally with support from the Data Office.

Whenever a project has an impact on data, all data aspects need to be reviewed before approval, through a project data review process. It needs to become a mandatory process, a precondition for all projects to obtain approval and/or funding.

The process describes criteria for this mandatory review as well as the steps of the review, the areas of assessment, and the parties in charge. The output is a data review document plus a summary that becomes a mandatory part of any project approval or funding request. The summary lists all potential deviations, possible options, and the final judgment.

Typical areas of review are CDM compliance, Single Source of Truth, and correct definitions and terms. The process would also determine areas where a project aims at changing the current CDM or verbiage, to ensure the change requests are submitted properly (see Data Logic Change Process later in this chapter).

Subject matter experts will need to judge initiatives from a data perspective – PowerPoint presentations to Executives do not guarantee architecture compliance.

For efficiency purposes, all of this can become part of a comprehensive Architecture Review, covering all Architecture aspects in one go. This is another good reason for any organization to have an Architecture Review Board (ARB) with members from the IT Architecture team and the Data Office.

Where the ARB does not agree with a project approach, the case needs to go to the Data Management Council, including different options and all known arguments for or against them.

The project data review process is probably the process that deals most frequently with Data Concessions: Not all solutions can realistically become data compliant in one go. Wherever there are good reasons to deviate temporarily, this may obtain conditional approval, together with a Data Concession limited in time.

You need a dedicated process for the steps to obtain a concession plus the management of existing concessions, including follow-up where concessions are about to expire or have done so. Before expiry, the solution must be compliant, or (only where there are good reasons) the concession needs to be extended, and the solution remains on the list of concessions to be monitored.

This process is not limited to waterfall projects, but agile projects require a slightly different approach as you cannot do a full review for each sprint. Instead, all foreseeable Architecture questions should be covered at an early stage, and only new findings during the initiative would trigger another review.

## When is a project relevant for data review?

Not all projects need to undergo a data review. As a general rule, a review is required in the following cases:

- A process is touched that deals with data (retrieve, create, modify, delete).
- Software is to be built or changed that deals with data (retrieve, create, modify, delete).
- Changes in business logic result in changes in data logic.

**Examples:**

- Where a data source is added to a data warehouse or where data from a data warehouse is used for operational purposes
- Where new software is introduced to maintain data locally (or in the cloud if cloud-based)
- Where an application needs a list of facility codes
- Where an application sends masterdata or transactional data to another application
- Where data gets mapped to be exchanged between different legal or geographical entities
- Where data needs to be validated during data entry or after data transfer
- Where data from different entities (e.g., acquired organizations) is merged into one single database
- Where data is intended to be duplicated to be made available to another application or user group
- Where data privacy is (potentially) impacted
- Where Data Quality is a critical factor

## Cases where Data Management is not the primary point of contact

The following cases may sound like data topics, but other functions should be in the lead:

- Physical data connections (LAN/WAN): To be checked with the IT Infrastructure team.

- Information security: This is a topic for the InfoSec team.

- Data center operation.

- Data transfer protocols.

- Additional hard disk space to cater for an increase in data volume.

- Implementation of a data encryption algorithm.

## A typical project data review

A team suggests a new solution where certain elements of Masterdata would be maintained in spreadsheets and distributed via email, or where terminology is to be introduced that is not aligned across functions.

- This will be made visible during the project data review process, and discussions with the project team about alternatives would take place.

- Whatever option the project finally decides to proceed with will be judged in a data review document first reviewed by your Data Management Council.

A summary will also be provided to become a mandatory part of the project approval/funding request.

# Support processes

Such processes are required by data teams to provide support to functions, particularly in preproject phases (right direction while still possible)

Finding out about noncompliant data handling during a project approval process is too late. Data experts need to get involved as soon as a project starts its scoping phase. This includes both Data Architects and Business Experts from Data Management.

These experts would help define a compliant approach. They would assist in creating different options and possibly phased approaches. They would regularly review the current status of thinking within the Data Management community.

# Data content change process

Changes to data content usually have limited impact, allowing for a lean process. Risks come where data handling is not compliant, e.g. where reference data has been hardcoded in an application so that a change in the reference data repository does not update the application.

Most data changes are already based on operational processes, usually defined by executing business functions. These processes need to fit together, and some processes may be improved based on experience with other processes. This is a long-term activity, for which a proper stocktaking is an important first step.

Eventually, changes to Masterdata should be a matter of configuration. Most organizations have not yet reached that level of sophistication. That is why such changes often require an impact assessment across various business functions. Such an assessment needs to be part of any of the existing functional data change processes.

## Data Quality Management process

The quality of data needs to be monitored, and deviations from quality targets need to be addressed.

Depending on the impact of quality issues, you need to react differently:

(i) **Accept**

Action: Do nothing, as the impact is too small (that is why it is important to quantify the impact of any reported issue).

(ii) **Mitigate**

Action: Apply workaround (e.g., manual processing), or implement automated preprocessing software – until integrated, or through a change of mapping logic (maybe accepting other impacts in reverse).

(iii) **Solve**

Action: Change software (may even be legacy software that will be retired in 2 years' time), or change a process, or change a business rule (or a combination of any of those).

## Data logic change process

You need different processes for changes to the data structure, data rules, data processes, or data glossary.

In order to avoid uncoordinated data activities, all requests for data-related changes need to follow this process. The most critical request is that for data model changes as a result of business logic changes.

The process ensures the intended changes are unambiguously described, and the impact on all other functions is assessed. It will suggest an enhanced project scope where additional activities are found to be necessary to avoid unwanted impacts.

A cross-functional approval flow and concession handling (as described earlier) need to be part of this process.

# Data glossary process

You can imagine that it is not sufficient to have a glossary and a person in charge.

In addition, you will need to develop a process that specifies whom to talk to in case of questions, proposals, or requests for clarification.

You will also need to define the parties required to review any such input, plus the steps leading to a decision or a response.

# Data access request process

Data access to data platforms requires a trade-off between the users' need for flexibility and the platform providers' interest in usage that is adequate and diligent.

Such a process should aim at reducing the number of users to the necessary minimum while giving those few users with the necessary qualification and business needs enough flexibility in using those platforms.

Criteria:

- Does the user have the skills?
- Is there a business justification?

# Manage data in business processes

Data topics are not only to be handled through dedicated data processes. In addition, there are data components with any of the existing business processes.

Your Data Governance function should govern those parts of the business processes for the following two reasons:

- You may want to ensure data is always dealt with according to your data principles, standards, and rules, even in departmental routine work that does not involve data specialists.

- Different business processes may share the same data needs. They should use the same data subprocesses to ensure consistency. Application development will then be able to have various business applications use the same APIs or web services, thus avoiding duplication on the IT side as well.

But how do you get from business processes to data?

You are in a lucky situation if your organization has an established process management function. The better an organization's processes are documented, the easier it will be to determine the data-related subprocesses.

Where this is not the case, you may use the data level to force proper documentation of processes. It is often a viable option to reengineer processes from the application workflow. If your Data Office supports a business team in this exercise, the determination of data subprocesses can become a welcome side effect.

You determine data subprocesses by analyzing the data flow for each step of a business process:

- Where is data input expected?
- Where is data to be looked up?
- Where is data to be validated (which usually includes lookup of data to be validated against)?
- Where is data to be modified, merged, deleted, and so on?

It is important to document all findings and results, both within the business process documentation and as part of the Data Office documentation. You also need to add cross-references to any web service or API used.

You find a (simplified) example of a data subprocess in Figure 8-2.

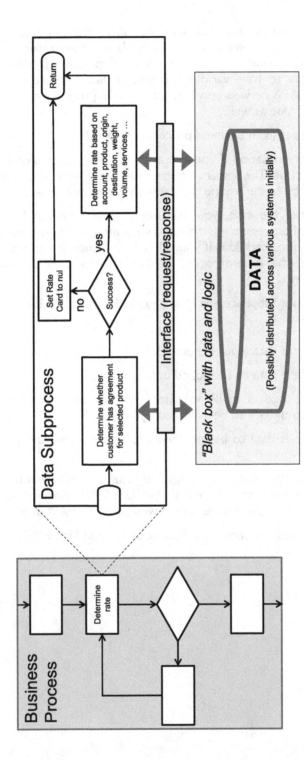

**Figure 8-2.** A data subprocess

Another motivation is that this approach decouples technical data handling from the process flow. The business process owner does not need to care about the inside of the data subprocess. It is sufficient to state WHAT is needed, and the HOW can be left to the data experts.

# Roles and Responsibilities

*"How about—bear with me here—we actually try
to change something?"*

**Figure 9-1.** Do you always know who does what?

© Martin Treder 2020
M. Treder, *The Chief Data Officer Management Handbook*,
https://doi.org/10.1007/978-1-4842-6115-6_9

# Introduction to roles and responsibilities

Why have we dealt with processes *before* looking at roles and responsibilities?

The answer is straightforward: Each role, each responsibility needs to serve a purpose. A role that does not appear in any process is not required.

You'd ideally start by deriving responsibilities from processes. Once all responsibilities are clear, you can shape roles that you associate with them. Here you consider which responsibilities should be with different roles, and which responsibilities belong to the same role. Have Segregation of Duties and the four-eyes principle guide you here.

A good starting point for responsibilities is the list of data-related processes which we discussed in Chapter 8.

Based on the tasks we had determined there, we can already distinguish between functional teams with content responsibilities, functional teams with maintenance responsibilities, and centralized data roles.

On the IT side, we can identify technical architects and application architects. Other IT people may play important roles in handling data as well, such as database designers and software developers. My personal recommendation, however, is to leave the internal IT organization to the CIO and base the collaboration with IT on deliverables.

# Data Owners and Data Champions

Data Management needs business input. You may ask Business folks to contribute to and influence the management of data in two ways: By representing a *part of the organization* (such as the Sales department) or a *data domain* (such as customer or product).

It is essential to distinguish between both roles as data domains are usually cross-functional.

No common term has been established for either role, so far. I decided to refer to the frequently used terms of Data Owner and Data Champion, as shown in Figure 9-2.

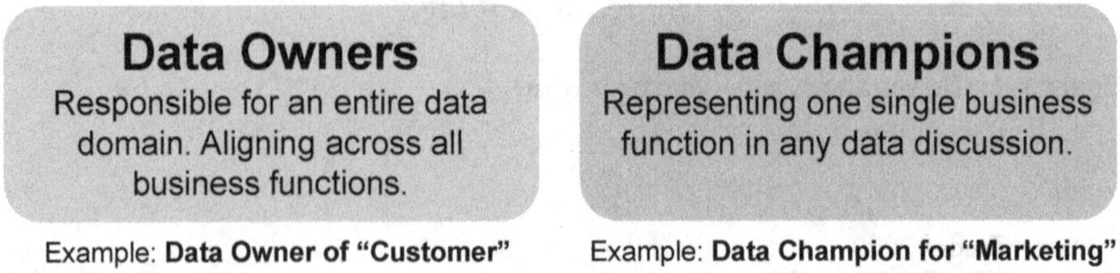

| **Data Owners** | **Data Champions** |
|---|---|
| Responsible for an entire data domain. Aligning across all business functions. | Representing one single business function in any data discussion. |
| Example: **Data Owner of "Customer"** | Example: **Data Champion for "Marketing"** |

**Figure 9-2.** Data Owners vs. Data Champions

Please feel free to replace those two terms with titles that match your organization's habits and culture.

# Data Owners

Data Owners are members of business teams who assume responsibility for a particular area of an organization's data universe from a content perspective, usually as a subset of the Corporate Data Model (CDM). Responsibilities of Data Owners must never overlap. Each part of the CDM must be unambiguously allocated to exactly one Data Owner. Please do not allow for a Marketing Customer Data Owner and a Finance Customer Data Owner to co-exist. The Data Owner of Customer has to align with all stakeholders.

# Data Champions

Data Champions can be considered the data advocates of a particular part of an organization.

We can distinguish between functional Data Champions (e.g., Data Champion for Finance), organizational Data Champions (e.g., Data Champion for a subsidiary), and geographical Data Champions (e.g., the Data Champion for Southeast Asia). Here you can see that responsibilities may indeed overlap, but the perspective is different.

Such Data Champions do not need to take a cross-functional perspective in the first place. You would ask the Head of each business area to appoint a Data Champion in whom he or she trusts, as the Data Champion will represent that business area in all data discussions of the entire organization.

Typical responsibilities of a Data Champion:

- Be the single point of contact toward the network, that is, for cross-functional data topics from the Headquarters and from all entities.

- Be the single point of contact for any data-related question, inquiry, or proposal from anybody within the represented entity or function. Share and explain data messages from the Headquarter. Ensure the organization's Data Principles are explained and understood within their own entity. (If you have Data Champions with overlapping areas of responsibilities, you should determine upfront whether, for instance, the regional or the functional structure is considered the primary one.)

- Local data network: Know the local Data Stewards so that they can be brought in contact with each other. Have them understand how their work impacts the work of Data Stewards in other functions.

- Ensure local expertise is brought into global discussions, for example, knowledge about local address handling habits, best sources of external data, or local legislation impacting data handling.

- Ensure local projects explicitly consider data aspects and support them from a data perspective. Collaborate with the central Data Office for a consistent approach.

- Have an overview of local controls to ensure Data Quality on a day-to-day basis.

- Observe local activities and look out for data-related issues. Run local initiatives and root cause analyses based on data findings, ideally using Six Sigma.

- Consolidate local needs with regard to data: Requirement toward tools (e.g., analytics visualization), data sources, country-specific data structures (e.g., for required trading licenses), validation of local data sources (e.g., in case of statutory changes).

- Determine best practices (as well as approaches which did **not** work well) through the network, in preparation of own or supported local projects.

- Facilitate the discussion where people from different functions have different views on how to handle data locally.

# Data Creators and Consumers

Data Creators and Data Consumers are two important stakeholder groups, but they represent entirely different roles. Please resist the temptation to address them together.

These roles are usually not full-time jobs. They describe sets of responsibilities assumed by people in business departments on top of their functional responsibilities.

**Data Creators** maintain data on a daily basis in their functional business roles. They are responsible for the content and quality of the data, but not for its structure.

**Data Stewards** are a subgroup of an organization's Data Creators. They are experienced in handling data, and they usually maintain a permanent dialogue with the Data Office and the Data Owners. Within their functional remit, they actively watch out for Data Quality issues and opportunities to improve Data Quality further. They are involved in the development of data metrics (see Chapter 10), and they can give experience-based input to the development of processes and applications from a data perspective.

**Data Consumers** are people from all business areas and of all professions that have in common that their work is predominantly dependent on data. They rely on good Data Quality, usually without being in a position to influence it immediately.

They know firsthand what it means to work with bad data. That is why they are an important voice in managing Data Quality.

The most dominant group of data consumers are the Analytics and Data Science communities. But there are many more consumers in operational processes, Finance, Customer Service, performance, and regulatory reporting, to name just a few.

You can already see that Data Creators and Data Consumers are not necessarily mutually exclusive groups. Roles that enrich data usually belong to both groups. People in such positions are generally those you can convince most easily that they need to take care of Data Quality.

The same applies to people at the end of the data supply chain, particularly Finance and Analytics. Whatever goes wrong anywhere along that data supply chain will impact them, inevitably.

But where these people understand the need for good Data Quality, they often tend to think they need to take care of Data Quality on their own – to fix the data before they use it. You find this behavior particularly in organizations without a tradition of active data management.

It is crucial for those people to first understand that fixing data just for a single purpose is not a sustainable approach. They should be encouraged to *report* any observed data issues instead, so that the organization can find the root causes and fix any issue where it occurs.

But you also have to demonstrate the Data Office's willingness and ability to manage Data Quality along the data supply chain. Data Consumers will only leave that work to the Data Office if they see success.

An actively managed Data Consumer community is a suitable means of conveying your messages, and it is a great feedback channel.

# Other business roles
## Business ownership roles

Business Application Owners and Business Process Owners are essential roles as well. And if executed properly, they will take a cross-functional perspective.

But please don't mix these roles with those of your data owners. They are equally important but separate, independent roles.

Mature organizations orchestrate a cross-functional dialogue between all of these roles, often under a Chief Transformation Officer (where the business model is dynamic) or a Head of Global Business Services (where the business model is considered mature). A Data Office should be part of this dialogue. The CDO does not need to be in the lead, though.

In organizations without such a systematic dialogue, you may wish to get active on your own. You might orchestrate the data-related roles in a way that makes the organization want to see a similar setup for all cross-functional topics.

In other words, somebody needs to take the first step – why shouldn't it be you?

# Lack of ownership

Where you don't find an owner, use the principle of "executive ownership": Inform the executive in charge that he or she is the ultimate owner. The executive will have no time and thus hate the idea. The good thing is that ownership can be delegated down. The executive will happily make use of this option.

Sometimes you don't even find a function that would be a natural business owner of a piece of software. This is often the case with legacy solutions that have been around for years. IT may have maintained them all the time, but nobody has ever claimed business ownership, and nobody knows who uses the application. But all of a sudden you might need to address data noncompliance in such an application.

A possible, admittedly bold approach: Claim business ownership for the Data Office. Publicly announce decommissioning. Find out who panics most. That person can become the owner to prevent uncontrolled switch-off. If nobody raises a hand, execute your decommissioning plan while broadly communicating every single step. And prepare a roll-back plan!

# Centralized roles

Following the subsidiarity principle (see Chapter **2**), certain roles need to be centralized to ensure consistent behavior and to avoid double parallel work.

Your Data Governance team should check each activity in any of the data-related processes for the need to be managed centrally.

Depending on the structure of your organization, the set of roles that would become part of a central organization will vary.

While you would determine an adequate list of central roles together with all other functions in business and IT, you should consider that these functions are usually biased: Most functional areas tend to prefer the centralization of all the hard work while being happy to accept the delegated role of decision-making.

That is why, ideally, you prepare a first organizational setup within the Data Office before you start the dialogue. This allows you to ask for good reasons whenever somebody proposes a change.

This chapter provides you with a list of groups of central responsibilities, broken down into possible teams. It is not intended to be exhaustive or prescriptive. It should rather help determine a good level of detail to start with.

And remember, the adequate number of teams or team members in your Data Office depends on the size of your organization, the budget of the Data Office and the degree of data literacy among the workforce (remember the tenth aspect of effective Data Management in Chapter 2).

The following structure is meant to be scalable in that a variable number of responsibilities can shape a single role and that a single team can cover one or many of the areas described.

# Data Governance

Here is a list of typical responsibilities of a Data Governance function:

- Define, maintain, and document **data maintenance** processes.

- Define, maintain, and document processes to change the **structure** of your data (in other words, govern the Corporate Data Model, while Data Architecture is responsible for its content).

- Manage **external data** sources and standards (including agreements with external data suppliers).

- Develop **training** and execute **communication**.
- Document and publish **data policies and standards**.
- Describe **data rules** and **logic** (with input from Data Architecture).
- Manage a single, cross-functional **glossary** of business terms.

# Data Quality

Hardly any other data discipline requires business knowledge as much as a good understanding of data and people's behavior. The following list of responsibilities of a DQ function makes no claim of being complete:

- Define and classify reliable metrics (**Data Quality Indicators** – see Chapter 10) and **reports**.
- Manage and measure **Data Quality**; use metrics and heuristics.
- Validate existing and proposed Key Performance Indicators.
- Work with **business functions and IT** toward a single view on Masterdata and transactional data.
- Establish and publish **dashboards** for Data Quality.
- Determine key issues and run **root cause analyses** (**Six Sigma** – see Chapter 11).

# Data solutions and projects

Data Management should not be limited to the roles of rule-maker, police, and helpdesk for data. Someone has to take data out of its ivory tower.

In other words, the expertise and knowledge about data need to be transitioned into daily processes and applications.

You will probably need to cover the following responsibilities in the area of data solutions and projects:

- Be the business owner of all primarily data-related software of cross-functional relevance (predominantly databases, data maintenance tools, and web services).
- Manage solution and implementation **road maps** (to be based on business priorities, technical capabilities, and own cost/benefit analyses – see Chapter 16).

- Coordinate **change requests** against data solutions cross-functionally, and lead the implementation project from a business perspective.

- Execute or organize **usability analyses**.

- Orchestrate global **networks** of Data Champions, Data Stewards, Analytics teams, and subject matter experts.

- Proactively support and guide **migration** and **transformation** projects in data matters. Deploy data applications owned by the Data Office, and support the deployment of other applications from a data handling perspective (e.g. in the area of country-specific masterdata configuration).

- Manage data-related noncompliance ("data debt").

- Organize **cross-functional** collaboration.

# Masterdata management

While *maintenance* of Masterdata usually takes place within the respective business functions, you will need to coordinate all of these activities.

Here are a few typically central Masterdata responsibilities:

- Ensure proper maintenance of **Masterdata, Reference Data**, and **Metadata**.

- Coordinate **impact analyses** of significant Masterdata changes.

- Monitor proper **usage** of Masterdata.

- Document the Masterdata universe.

- Be the business owner or product owner of MDM tools.

# Data Architecture

Data cannot be architected within single business functions, nor should it be done by technical IT architects.

The reason is that, as with buildings, you need one single overarching architecture to reflect your organization. Content input must come from the various business functions, but all of it needs to be cast into one single Corporate Data Model.

There is no objective delineation between Data Architecture and Data Governance in the area of data standards. Some organizations distinguish between organizational and functional standards: While organizational standards are with Data Governance, functional standards are with Data Architecture.

While there is no right or wrong, I personally recommend that the entire responsibility for standards be with Data Governance, with functional expertise to come from Data Architecture. This setup avoids uncertainty about the responsibility of borderline cases while still capitalizing on Data Architecture expertise.

Following this setup, typical tasks of a central Data Architecture team comprise

- Maintain a Business **Data Model** based on business processes and capabilities
- Perform **gap analyses** on the Data Model
- Support Data Governance in shaping and enforcing **functional data standards**
- Align with other Architecture disciplines
- **Assess the impact** of business changes on data structures

## Data privacy and compliance

These topics are older than the concept of a Data Office. That is why most organizations already have an owner. And, honestly, you might not necessarily have to strive for ownership here. Eventually, from an organization's perspective, the key objective is to have proper ownership in place. Where this is already the case, there is no problem to solve, so don't try to.

Instead, you need to ensure the Data Office is accepted as a stakeholder here. In many organizations, the Legal department has a perfect understanding of what compliance means – but they may not be able to translate it to practical guidelines for daily work.

This is where a Data Office can offer its support, including the provision of concrete examples and check lists.

Note: Where no other department assumes ownership, or where the Board might want the Data Office to be responsible, accept it! A Data Office is not the worst place to put the responsibility for data privacy. But in this case, it should be you who reaches out to other departments for their advice, for example, in legal matters.

And here is a typical list of Data Office responsibilities around data privacy:

- Work with the Legal department and with Information Security on **compliance** and **data privacy**.

- Support business functions in applying data privacy regulations correctly.

- Ensure data privacy policies and processes are target-aimed, as unambiguous as possible and understood.

# Data Science

In most organizations, it might not make sense to centralize all Data Science activities – you might lose the necessary proximity between the Data Scientists and the business functions they are working for.

However, it is valuable to build a small yet strong central Data Science team as part of the Data Office. Ideally, you shape a "Center of Excellence" for Data Science as the scientific hub of your organization's Data Scientist community – something that functional Data Scientists are often not given the opportunity to do.

Typical responsibilities of such a Center of Excellence are

- Evaluate and **share the knowledge** of Data Scientists across all business functions

- Represent your organization to the **outside world** in the area of Data Science (which does NOT mean blocking other Data Scientists' external contacts!)

- Run cross-functional **data science** activities (correlations, causalities, probabilities, forecast)

- Work with IT on a Target **Analytics** landscape

- Exploit younger data concepts such as Blockchain or the Internet of Things (and involve functional teams where possible)

# Data Analytics and BI

Again, a lot of Data Analytics and Business Intelligence (BI) should be done within the business functions, the more so as concepts such as BI self-service make it easier even for semi-experts to turn data into information.

However, a Center of Excellence can add value here as well, as a coordinating function and by helping business functions using data, tools, and algorithms accurately. It is important to make friends with those business functions and to avoid being perceived as competition.

Here is a list of responsibilities you would find with an Analytics and BI Center of Excellence:

- Support functional **Analytics and BI team** from a technical perspective
- Represent your organization to the **outside world** in the areas of Data Analytics and BI
- Work with Data Science and IT on our Target **Analytics** and **Data Visualization** landscape
- Run cross-functional **Predictive Analytics** projects with business functions
- Support Business **Reporting**
- Work on **Big Data** concepts: What is new in handling and leveraging external and internal mass data?

# Data Quality

*"Yes sir, you can absolutely trust those numbers"*

**Figure 10-1.** Data Quality? Cross your fingers and hope…

© Martin Treder 2020
M. Treder, *The Chief Data Officer Management Handbook,*
https://doi.org/10.1007/978-1-4842-6115-6_10

# Why is Data Quality important?

"Data Quality" *sounds* important. And hardly anybody would question the importance of Data Quality. That is why it has the potential to become another hollow catchphrase. You appear really professional if you expect "high Data Quality."

And, yes, Data Quality is indeed important. Data Quality is so important that I am using a dedicated acronym. To me, it is simply DQ, and DQ should be at the top of your list of priorities.

Unfortunately, some people think that good Data Quality is the norm as long as nothing strange happens.

Good Data Quality is not the norm. Not at all.

In other words, unless an organization works on Data Quality, actively and proactively, you can assume this organization's Data Quality to be bad.

---

## EXAMPLE I

An internationally active company recognizes the need to cleanse address data in certain country organizations.

This company decides to task a huge team in its head office with the cleansing work.

The team failed.

Eventually, the data had become syntactically correct. But it could not be used to create a Customer 360 view, and it was not even suited for the allocation of customers to sales areas.

The problem turned out to be manifold:

- Incomplete capturing of customer data for years resulted in lack of attributes that could not be derived from the other data.

- Furthermore, country organizations had been allowed to misuse some of the data fields for country-specific purposes so that the content was useless in an overall corporate context.[1]

- Finally, cleansing customer data required a lot of local knowledge about the habits of providing addresses and about company structures in each country. The team in the head office did not have that knowledge.

---

[1] David Millan, Global Head of Data at Unilever, stated at a data conference in 2019 that, in his organization, 80 percent of allegedly country-specific points had turned out to be universal.

# Dangerous Data Quality standpoints

In organizations without a dedicated focus on Data Quality, I have found a number of different patterns which I am describing here.

Feel free to check your own organization against these patterns. Awareness allows you to tailor your DQ strategy accordingly.

## Assuming your DQ is good

Why do organizations often focus on Analytics, instead of covering the entire data supply chain?

Very often, it is because their deciders think their data is in good shape as a matter of course ("Well-managed organizations have good data!").

This coincides with a prevalent, subconscious assumption that all of their data is good, as nobody has complained so far.

Sometimes the leaders of an organization haven't even thought about DQ. And I understand them well! Think about it: When have you consciously thought about your ears the last time? It has been a while? Well, this is normal! People only think about their ears when they have earaches or start to hear badly.

It is the same phenomenon we observe around DQ. It is not part of an executive's daily thought process.

That means we don't face an intellectual issue but an **awareness** issue.

I recommend that you address this problem first, well before you suggest any technical or organizational measures to improve your Data Quality. It is hard to work on DQ (let alone to get funding for such work) if there is no executive support.

## Assuming your DQ is good enough

You may have heard the following statement every now and then:

> It may not be perfect, but it has been sufficiently good so far. Let's follow the 80/20 rule, and let's not exaggerate!

But how to tell whether 20 percent of the effort really leads to a quality level of 80 percent? And even if it does: Are 80 percent really sufficient?

## Assuming you cannot measure DQ

The fatalists say:

> You can only measure DQ against other data – which may be wrong as well.
> So you never know!

Yes, DQ is not easy. It requires complex planning and determination of dependencies. But it is possible to measure data – at least to a certain extent. Plausibility checks work in most situations. And as long as you are aware of the limitations, any measurement is better than not measuring at all.

## Assuming bad DQ is a "Data Office" task

Who is supposed to fix bad data? The data experts from the Data Office?

Insufficient Data Quality is often **not** a problem of the data itself but a business issue. In such a case, data experts can help describe the problem and the impact. They may even support in fixing the data, but they should not fix the data on their own, for three reasons:

  (i) **The content knowledge that is required to fix data should be with the business owner.**

  (ii) **Business people should be forced to take ownership –** *their* **data is a core part of** *their* **business.**

  (iii) **Without addressing the root cause, Data Quality will quickly deteriorate again.**

Thank goodness, there are ways for business functions to work on the quality of their data. They can use Six Sigma to determine the root cause so that they don't just cure the symptoms, and they can apply proven Six Sigma methodologies to address those root causes.

I encourage you to share the Six Sigma approach with all business functions. If Six Sigma is already part of your organization's toolbox, support those teams in applying it to data.

See Chapter 11 for details on how to use Six Sigma in the context of managing data.

## Assuming everybody wants good DQ

Let's be honest: Some people in an organization benefit from bad Data Quality. Without active DQ management and proper incentivization, they won't see any reason to contribute to good DQ.

As an example, think of the classic underperformers who don't want their performance to become visible through data.

But there are also those daily issues where employees benefit from bad data.

---

## EXAMPLE 2

Imagine an employee who is responsible for customer data entry, supported by a system that validates the format of certain data fields based on metadata.

Being rewarded for fast data entry, this employee faces a conflict between the number of customer records entered per hour and the accuracy of the data entered: You cannot spend too much time on single records to achieve your hourly target.

But sometimes data is missing or inconsistent, and it costs a lot of time to find out, for example, to check the customer's website to determine the VAT number.

As a quick solution, this person decides to enter syntactically valid dummy data in such cases. Result: Performance target reached – customer data wrong!

Good for this person – bad for the organization.

---

## Addressing DQ only once you are in trouble

It is easy to ignore DQ as long as it does not hurt you visibly. We had determined this approach as one of eight typical behavioral patterns – see Chapter 1.

An organization that deals with DQ reactively cannot avoid the impact. It can only delay it.

Some information may be lost for good if not taken care of during its early stages. In such cases, it is simply impossible to regain good Data Quality later.

If an organization without active DQ management wants to base a decision on data, it may face the following issues:

- The data is incomplete or not available at all because there was no plan to prepare it for decision-making.

- The data is ambiguous. People can interpret it in different ways, depending on their personal or departmental objectives.

- It takes a long time to gather and fix all the data. As a result, decision-making gets delayed.

**Figure 10-2.** Do you have reliable, consistent data?

## Working on DQ for Analytics purposes only

During a big conference, an expert stressed the importance of Data Quality. His demand: "Ensure the data in your data lake is well-maintained."

A clean data lake is good. But starting there with Data Quality is too late!

Unfortunately, the necessity to obtain reliable analytics results is often the only driver for Data Quality. This is one of the weaknesses of organizations that consider Data Management to be identical to Analytics. If they have a DQ team, it is part of their Analytics department.

Such Analytics-driven Data Quality usually doesn't address the root causes. Instead, it tries to fix the data in the data lake – or even later.

And this perspective makes Data Quality reactive: You try to fix it once you find out that it is not good enough.

This approach comes with further downsides:

- It creates a false sense of Data Quality: It seems to be a nonissue as long as the Analytics people do not complain.

- You often cannot tell from data in a data lake whether it is of good quality – particularly if it is syntactically correct.

- Not only may the DQ issues in a data lake go unnoticed – the incorrectness of subsequent analyses may escape notice as well.

- People may not focus on avoiding Data Quality issues in their daily lives if "those Analytics folks fix it before they use it."

- In any organization, data is not only used at the end of its journey. It goes through several stages before it may finally end up in a data lake. A lot of data is used for operational purposes as well, such as the creation of invoices or the steering of production processes. Such activities need high-quality data as well. Nonanalytical data users that take their data from the operational data repositories do not benefit from any data fixes in the data lake.

- Working on Data Quality for analytics creates inconsistencies between the (uncleansed) data used for operational purposes and the (cleansed) data used for analytics.

- Data repaired late may already have caused damage or resulted in wrong messages before.

## EXAMPLE 3

Data entered on a website by a customer may be used to take further decisions during an online booking process. That means that data is already used before it is even transmitted to the back end.

It doesn't help the booking process if the data gets fixed later, for example, through an address validation process.

# Working on DQ where first problems arise

Even where DQ is a topic beyond Analytics, people often fix data where problems become **visible**. But things may already have gone wrong earlier. Such organizations typically face the following issues:

- Once certain aspects of data are lost, it may no longer be possible to recover them. You cannot, for instance, derive a customer's satisfaction rating from the remaining details of an online purchase.

- If you repair data along the way, you might end up with inconsistencies between "before" and "after" the fix. Processes that use the unfixed data will not match those using fixed data.

- Processes that are responsible for bad data are not addressed. Fixing bad data becomes a recurring, retrospective task.

Data Quality needs to include the prevention of data losses and timely activities to recover. If a data import or data entry is found to be insufficient, the software can still react and request a near-time closure of the gap.

## Accepting bad DQ because better DQ is impossible or difficult

Very often, particularly data scientists need a certain type of data for a project. They may search the Web and finally find a CSV file somewhere, or they scrape a table from a website.

This is not an invalid approach as such, even if it is often possible to get the same data at better DQ elsewhere. In most cases, however, the data scientist could at least fully understand the level of DQ of any such data repository.

And I am not speaking of context-free assessment of a given file.

Instead, the data scientist would also focus on the source: How reliable, how old, how unbiased?

This has even become a discipline on its own, often called EAI – Exploratory Artificial Intelligence.

The discovery of bias is one of the important tasks in assessing DQ. And this is not primarily about determining malicious falsification of data. Cases where public institutions systematically share manipulated data files are certainly the exception. But bias already starts where somebody has filtered out the perceived "outliers" before you get to see the file and without you knowing the filtering criteria.

This is an example of the need for active DQ management: Each data scientist can mitigate the risk by searching across multiple sources. But this is additional work, which would not be required to successfully apply a Machine Learning algorithm.

That is why data scientists must be encouraged and rewarded to expend the necessary additional effort.

# Not communicating the level of DQ

Sometimes data is not perfect – which does not necessarily render it useless. Bad DQ certainly limits the significance of the results, but in many cases, those results are more valuable than no results at all.

However, in order not to overinterpret the results, you need to provide the level of DQ together with those results.

It is often a good approach to work with thresholds based on worst-case scenarios. A typical, valid statement could be

*According to our model, the expected value of the KPI is 80 percent. But we have reasons to assume bad DQ: (Add reasons here.) The resulting confidence interval is as big as 15 percentage points so that we can assume the true value to be between 65 and 95 percent, at a probability of .95.*

# How to deal with Data Quality?

## DQ must be a top management topic

Data Quality has an impact on all other kinds of quality, and it influences all information provided to the Board.

That is why the Board should be interested in the quality of their organization's data. They can be informed through a selected number of consolidated metrics provided by the (unbiased, neutral) Data Office.

## DQ requires the right motivation

When managing DQ, carrot and stick need to be balanced. You will not achieve data quality if you simply penalize providers of bad data.

In essence, people need to be given good reasons to deliver great data quality, particularly where they are not consumers of the data they provide.

A balanced approach would be somewhere in between a penalty-based system that will make people try to hide bad quality and an inflexible bonus system that runs the risk of rewarding the wrong behavior.

Two ways of encouraging DQ are transparency and support:

(i) **Transparency**

Sustainable data quality can more easily be achieved by making excellent work visible.

Eventually, regular communication of DQIs can trigger a kind of "best data quality" competition. That is how you want people to think.

(ii) **Support**

Another important capability of DQ is to add advice.

"Your quality is low" is not fostering the right DQ mindset. Saying "Here are a few ideas for you to reach your DQ targets in future" helps increase acceptance of the messages that someone's DQ is insufficient.

## Let the right parties raise their hands

Don't fight for high data quality on your own account. Remember, it is not the Data Office that requires good data for its daily work.

Instead, you may want to determine the business functions that require high data quality or at least benefit from it. Make **them** ask for better data!

You may need to explain the impact of bad data first – many business functions may not be aware. And then give them a voice – act as their advocate.

## Focus on relevant data

Data may be correct but useless. Checking your data for relevance is, therefore, part of proper data quality management as well.

The relevance of data often depends on the purpose more than on the data itself.

At times you may be tempted to run a regression analysis to find out possible causality between some attributes and an outcome – but common sense already tells you that, for some of these attributes, there is none. If you still go ahead, your model may suggest causality because it determines correlation.

You can reduce the risk of false positives if you exclude such data attributes from your analysis right away. In this sense, Exploratory Data Analysis (EDA; see also Chapter 20) is a critical aspect of data quality in Analytics.

## Keep and get data clean

While proper data management would focus on **keeping** data clean, data **cleansing** experts will always be required. Old buckets of data will always pop up, requiring cleanup. And even clean data ages over time, often requiring cleansing from time to time.

Especially where you see data come from an external source or a specific geographical area, you will need specific data knowledge for efficient cleansing.

This is a good reason not to limit data cleansing to a central team, no matter how intelligent those experts may be.

---

# EXAMPLE 4

The fight against duplicates requires more than a string comparison. That is why avoidance of duplicates and deduplication go beyond database operations. They also require business knowledge.

Instead of asking "Are records identical, or sufficiently close to be assumed as identical?", you may wish to ask "Are records referring to the same thing?"

If the answer is "yes," they are duplicates, no matter how different they look.

As a concrete case, compare these five location records in the UK:

1.   W1K 7TN

2.   86-90 Park Ln, Mayfair, London

3.   Hotel JW Marriott, Grosvenor House, London

4.   51°30'36.1"N 0°09'15.8"W

5.   ///ahead.foster.waddle[2]

Each of these five records unambiguously identifies a single location in London, Great Britain. In fact, they point to the same place. That means they are duplicates. You need some background information to find out and to join them into one record.

And you might wish to keep details from each of the four records so that the final record contains more information. What is the benefit here? You can check new records for congruence with such an enriched record more easily, and you could support intelligent data entry, by offering this record after a user has typed in the first few characters of any of the different attributes.

Note, however, that you will need to incorporate your data model to indicate which identifiers are unique on their own and which of them are unique in combination with others only: While, in the preceding example, the British postcode is exclusive to this place, the geo-coordinates might equally apply to each floor of a skyscraper (making it necessary to add a vertical Z coordinate for uniqueness), and London has two different buildings called "Grosvenor House."

---

[2]What3Words identifies all locations on earth by three words taken from a normal dictionary. See https://w3w.co/ahead.foster.waddle

But how do you implement proper data cleansing, using both human knowledge and computer knowledge?

(i) **Use data and algorithms.**

Thank goodness, a lot of knowledge has meanwhile been incorporated into algorithms. Professional address management solutions would already suggest that all records from the preceding example point to the same place.

That is why available computer knowledge (mostly external APIs which guarantee automated maintenance of data and algorithms) should be used in a first step.

(ii) **Use human experts.**

But you need human validation as a second step, in all cases with insufficient confidence in the outcome of the algorithm.

You may, for instance, ask colleagues from a particular country to help validate and cleanse customer data from that country. Those colleagues may have a better understanding of the country's address structure or legal aspects than highly skilled central resources.

(iii) **Validate as early as possible.**

If you import external data or merge data from an acquired organization into your data repositories, this cleansing step should take place early during that process. Asking local specialists to tidy up afterward may be too late – you might already have lost alleged duplicates for good in an early technical validation step. The rule of thumb is "cleanse before transforming."

So, instead of the traditional

[extract]→[transform]→[load]→[validate]

you can enhance the process to become

[extract]→**[validate]**→**[cleanse]**

　　　　　　　→[transform]→[load]→[validate]

## Everybody should be responsible

More and more data scientists ask for high-quality data. But is it sufficient to regularly tidy up your data lake?

No, it is not. As mentioned earlier, you need to manage data along the entire data supply chain. The best place to apply Data Quality is upon entry. This prevents bad data from making it into the system.

If an organization makes Data Quality part of each role and process that deals with data, it will achieve more sustainable Data Quality. Moreover, the overall effort will go down, as it is easier to keep data clean than to clean it later.

## DQ needs to get measured

How do you find out that your Data Quality is good? How do you determine any improvement over time?

It is by introducing metrics and heuristics and by institutionalizing their permanent application throughout your organization.

In a second step, you will need to define what is "good enough" and which level of DQ corresponds with which business impact.

This helps you measure DQ "against something" to determine any need for action.

We will get to the details later in this chapter.

## DQ needs to lead to action

It is not sufficient to find out that Data Quality is bad. Something must be done about it.

Fortunately, most people will agree.

Unfortunately, people tend to move straight to a solution to fix the issue.

| EXAMPLE 5 |
| --- |

Imagine an organization that has a problem with duplicates in its Sales Leads database.

If you find out, you could quickly establish a deduplication process, potentially based on external data to fix misspellings of company names.

At second glance, you may wish to understand **why** you find so many duplicates among your sales leads.

I once faced such a case. I asked the team to do a root cause analysis to determine the reasons. We quickly realized that the organization's sales organization had set up a scheme to incentivize its salesforce on the generation of sales leads.

Now imagine a salesperson that finds a potential customer and enters it into the system. A message pops up, stating "sales lead already registered."

What is the salesperson tempted to do? Yes, he or she may consider changing the spelling slightly, for example, turning organization ABC into organization A.B.C.

What are the results? Bonus payments based on wrong information, duplicates with slightly different spellings in the database, misleading calculation of KPIs such as the conversion rate of sales leads, and maybe even a potential customer approached twice, independently, as the system generates two sales plans.

The problem to solve here is not primarily the removal of duplicates from the database (=curing the symptoms) but a change of behavior, possibly through a modified incentivization scheme.

In other words, you would address the **root cause**.

# Management of business metrics

Meaningful business quality metrics require active management of Data Quality. Why?

There is a broad consensus that Key Performance Indicators (KPIs) are important to steer the business. People tend to forget that KPIs are based on data. Bad Data Quality leads to useless (or even misleading) quality metrics.

That means before we can manage our business, we need to manage our data so that we know how to steer our business.

A key aspect of managing data is the definition of *Data Quality Indicators (DQIs)*. These DQIs need to tell you how good the quality of our data is.

It is important in this context that we also need to measure data *handling* quality, not only the quality of the resulting data.

But how do you use DQIs effectively? Here are a few thoughts.

## Measure the performance of teams

You can tailor your DQIs to measure the performance of concrete roles or teams.

Just as many organizations do with their KPIs, you can define hierarchies of DQIs, following organizational hierarchies. Several DQIs on one hierarchy level are consolidated into one DQI on the next higher level – which, together with other, equally consolidated DQIs, form again a combined DQI one level higher.

This approach has three advantages:

- Each organizational hierarchy level has a manageable number of DQIs to look at – starting with team leads, for example, in Sales, Finance, or Customer Service, and reaching up to the Board where the CFO might be held responsible for a DQI called "Finance Data Quality."

- The hierarchical structure allows for digging deeper where a consolidated DQI indicates an issue.

- The responsibilities for DQ are unambiguously clear.

## Measure the consistency of data

Consistency checks should not wait for a cause. They should be run as a scheduled activity – systematically and regularly.

And you might want to run them right before any significant change to the application landscape. A new algorithm may not be as fault-tolerant as its predecessor so that previously undetected data inconsistencies may kill an improved algorithm.

## Consider heuristics

It is allowed to remove one postcode from a country's list of valid postcodes – the village may have been abandoned. But if this valid activity is done for half of the postcodes of a country, an alarm bell should ring.

In other words, doing a valid activity multiple times may not be legitimate anymore. But it might not be invalid, either. An algorithm simply does not have enough information to decide.

That is where heuristics come into play. Instead of letting an algorithm determine and act on cases of bad data entirely on its own, the algorithm would search for suspicious cases and report them to a human.

This approach aims at reducing the number of cases to be looked at by a human to a manageable number. That is also why such an algorithm would rather report many false positives than falsely accepting an error.

Humans would then decide whether a reported case is indeed an error, whether it is a false positive, or whether further assessment is necessary to find out.

Besides, such results would be used to fine-tune the algorithm, possibly with the support of AI.

## Determine unwanted behavior

Accidental encouragement of unwanted behavior happens easily, often through the design of general incentivization or through the DQIs themselves.

That is why you should keep an eye on the design and usage of all DQIs: Do they encourage or reward unwanted behavior? Do they still reflect business targets? Is the measurement logic still valid? Do DQIs need to be added?

As a general proactive measure, you might foresee a regular DQI review together with a defined, cross-functional stakeholder group. The Data Office should never define DQIs in isolation.

## Break down your quality measurement

Wherever you can, break down your quality measurement into distinctive groups!

This may allow you to determine the best-performing group and find out what they do differently – as long as the size of the groups is big enough for statistically relevant conclusions.

Sometimes it is not known which level of quality is "good" or "good enough." But you can always determine the trend: Does the DQI improve over time? What is the impact of actions on certain DQIs?

Here's my list of aspects to be covered by Data Quality Indicators:

a) Classical data input errors – through a user interface or mass upload

b) Data dissemination and usage issues

c) Data integrity issues

d) Structural data issues (errors resulting from correct data being stored under a wrong data model, e.g., a primary key violation may lead to the rejection of a record while, in fact, the key definition in the database table is wrong)

e) Data duplication

f) Completeness of data (even if all records are correct, incompleteness may provide wrong proportions)

g) Data aging issues (records that were once right but do not properly reflect changing reality anymore)

h) Duration of changes (data changes shouldn't take too long. Otherwise data is outdated or integrity at risk)

    i)    The percentage of problems or errors that got fixed

    j)    Timeliness of updates

    k)    Trends: How do DQIs develop over time?

    l)    Plausibility checks: Situations that are syntactically valid but "suspicious"

Breaking down quality measurement is also essential to gain Board-level attention for DQIs.

We all know that hardly any Board member would take the time to review a spreadsheet full of KPIs. Executives want to manage by exception, spotting the critical areas and focusing on fixing them.

As with most other metrics, it is, therefore, advisable to create a hierarchy of DQIs. This hierarchy needs to match the organizational hierarchy of the organization so that each DQI is owned by a human being, as the owner of a concrete role:

- Any role owner at any level is responsible for several related DQIs.

- These DQIs are usually consolidated into a summary DQI, representing that role owner's Data Quality level. The selection and weighting of DQIs are proposed by the role owner and approved by the role owner's line manager.

- As long as such a DQI is green (i.e., reaching at least a pre-agreed threshold), nobody else would ask the role owner to dig deeper.

- That person, however, would manage all DQIs on the next level of granularity, to take preventive action before one of them impacts the DQI they are measured against.

- In organizations where bonuses are based on managers' individual performance, such a DQI can form part of the calculation.

## Quality management to cover data

Is your organization's quality management certified? For example, by ISO 29000 et seqq? It probably is!

But has your Data Quality management explicitly been made part of that quality management system and certification? Do internal quality audits of your organization cover the quality of data?

Most organizations have not yet made this critical step. But you should use the unique opportunity, by seeking the dialogue with your organization's quality management teams.

They might be surprised to find allies from an unexpected direction so that they will probably be happy to join forces. And they should understand how the quality of service, quality of products, and quality of compliance are linked to the quality of the underlying data.

If you jointly work on enhancing the organization-wide quality management to cover data as well, you should use the opportunity and establish a genuinely cross-functional setup.

And if you manage to have DQ assessment added to the existing quality and compliance audits, you benefit in multiple senses:

- You use the existing authority of the quality management organization.

- You can reuse an established organization and approach.

- Business quality and Data Quality are looked at together – reflecting the close causal relationship between the two.

- You don't need to have your  busy data experts spend their time on Data Quality audits.

- Informally: You avoid being the bad guys. While you made the DQ rules, others help you enforce them – which does not always go hand in hand with making friends.

# Shaping Data Office Teams

*A Data Scientist is...*

*A Business Analyst that lives in California.*

**Figure 11-1.** What is a Data Scientist?

© Martin Treder 2020
M. Treder, *The Chief Data Officer Management Handbook*,
https://doi.org/10.1007/978-1-4842-6115-6_11

# The effective creation of data teams

To be effective, you need to design and set up your Data Office teams carefully.

Your teams need to cover all relevant data topics unambiguously, their roles need to be understandable to the outside world, and they need to be set up to collaborate effectively with each other.

There is no single best possible organizational structure. And please do not expect me to propose a generic org chart. Instead, I'd like to suggest a few sets of responsibilities that should be present in any Data Office organization.

# Data Architecture and glossary
## The "data language" team

As discussed earlier, the topics of a data glossary, rules, standards, and the data model are closely linked. A lot of interaction is required between these four areas.

That is why it makes sense even in larger organizations to keep all of them under the same roof. Such a department would cover the "logical" aspects of Data Management.

Depending on the size of your organization, you can break this department down into subteams. A typical structure consists of dedicated teams for Glossary Management, Data Standards, and Data Architecture.

## Glossary Management

Who should be in charge of managing your organization's glossary? I suggest you pick the least "nerdy" member of the entire Data Office team.

This person needs to have excellent interpersonal and communication skills as it will be in permanent dialogue with all business functions. Furthermore, it should be a language professional with a good feeling for the shades of a language.

Multilingualism is a plus. The primary purpose may not be to create a glossary in multiple languages but to be able to understand that there are numerous ways of saying the same thing. Eventually, data is "yet another language."

## How to organize Data Architecture?

The responsibility for the Corporate Data Model (CDM) should be with a dedicated Data Architecture team.

Some organizations have an Architecture team within their IT department. They often assume that all aspects of Architecture are to be covered by IT.

For good reasons, this is not the prevailing opinion among experts.

Let's have a look at a leading Architecture model, TOGAF.[1] It divides Architecture into four categories:

- Business Architecture
- Data Architecture
- Solution Architecture
- Technology Architecture

Business Architecture is a natural business responsibility, as the name already indicates. Application Architecture and Technology Architecture are classic IT responsibilities.

Data Architecture is the **bridge** between the two worlds, and this is precisely how a Data Office should position itself in an organization.

There might be some discussion as to where Data Architecture ends and where Solution Architecture starts, most notably in the area of databases and interfaces. In case of doubt, an organization would apply the following criteria: Anything related to **logic** sits on the Data Office side, while **technical** aspects reside on the IT side.

Depending on the complexity of an organization's data landscape, it might make sense to split Data Architecture into a logical area (Business Data Architecture) and a technical area (Technical Data Architecture).

The logical team, reporting into the CDO, is responsible for the business aspects of data architecture. It defines such diverse topics as

- The structure of the data (notably the Corporate Data Model)
- Which levels of freedom the Data Scientists have with regard to the Data Model
- What data must be provided to whom in which degree of predigestion
- Which parameters should be offered as part of a specific API

---

[1] The TOGAF Standard is an Enterprise Architecture methodology and framework provided by The Open Group. See TOGAF (2019).

Technical Data Architecture would consequently define the type of API, the technical data model, and the resulting table structure within the database.

Figure 11-2 provides a brief overview of the disciplines of architecture and where they sit in the organization. Note that, in the case of DevOps, IT operations-related questions are often dealt with by Application Architecture.

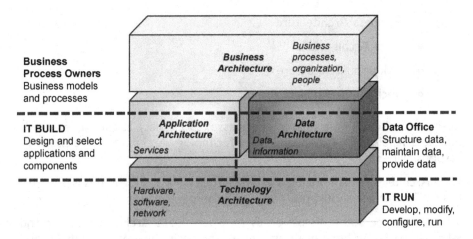

**Figure 11-2.** The TOGAF model

Despite possibly sitting in different parts of the organization, all Architecture teams should jointly work toward a single corporate architecture view.

This approach creates one single virtual team while having team members report into different parts of the organization, thus mitigating bias and silo thinking.

Such a joint working mode should be supported by a council with representatives from all Architecture disciplines. Such an **Architecture Review Board** must be empowered to jointly take decisions or to submit Architecture statements into an overarching decision process.

## How to make Data Architecture attractive?

Data Architects have a challenging role in an organization.

They tend to be considered the bad guys who primarily tell the business functions and project teams what **not** to do. As a consequence, those people often do not get involved voluntarily.

Unfortunately, good Data Architecture is like good security – in that its benefits usually remain invisible. It is bad or missing data architecture that has a visible, negative impact.

---

### ANALOGY

Why are Data Architects often less welcome in project teams than other Architects?

Project teams often perceive Architecture like this:

Technical Architects pave the roads, Solution Architects build the cars, Business Architects teach them to drive – and Data Architects tell them about red lights and speed limits...

---

Unfortunately, Data Architects are usually good at their subject matter, but, just as many other subject matter experts, they are often bad at self-marketing.

I remember a Data Architecture conference where dozens of seasoned Data Architects were complaining about the fact that solutions architects get invited to projects, but data architects don't. A quick survey showed that, as a consequence, the average ratio of Solution Architects to Data Architects in project teams exceeds 20.

No question: You have to help your Data Architects. You need to promote their value. Just making their involvement mandatory will not suffice.

Don't wait for people to understand the need over time. They will not. That is, they will not attribute any successes to Data Architecture.

In fact, next to Data Governance, Data Architecture is the area with the strongest need for promotion and explanation within the organization, to gain the necessary (and deserved) acceptance.

What can you do?

- Come up with real-life examples – from the past or from other organizations.
- Describe scenarios: What is going to happen where Data Architects do not get involved?
- Work with analogies (to take the preceding analogy further: Obeying both red lights and speed limits in traffic is actually good for your health).

- Stress the supportive aspects of your Data Architects' work – after all, they help projects find a feasible data setup, and they help avoid errors that could delay the project.

- Use your authority as the CDO. If *you* say Data Architects are crucial, it must be true.

- Teach your Data Architects to promote themselves (sometimes an external training does miracles).

# Analytics

The setup of an organization for the exploitation of data is complex enough for a dedicated book, and the best setup depends tremendously on the specific situation, such as the organization, its history, its organizational setup, and the amount of talent available.

A few considerations, however, may help shape such an organization, starting with the question of whether one single Analytics team for the entire organization is really the best choice.

## Analytics across silos

How can you run Analytics "as one"?

If you leave Analytics to IT, you will end up with a first-class race car without a driver: All the technology would be there, but hardly anybody could apply it to real-world questions.

Where Analytics is left to Business, however, you'd probably see various functional silos develop, especially in larger organizations. I have never seen a self-organized, cross-functional Analytics approach take shape successfully in such an organization.

Instead, you can expect each Analytics silo to develop independently. They will have experts familiar with their business area, which allows for the right questions to be asked.

On the other hand, the technical solutions will probably be second class as the functional Analytics department will mostly lack the critical mass to mimic an organization's entire IT intelligence.

Furthermore, a lot of business topics will be addressed several times in parallel, as those Analytics silos may not talk to each other. You see this frequently in organizations that are too big for one central management team. They subdivide management either into functional groups or geographical groups.

Federation is generally seen as an organizational necessity. It is well known that it does not make sense to regularly gather dozens of managers around the same table: You'd quickly see a small group discussing topics that are specific to a business function or a country organization, while the rest would get bored.

A federated approach in Analytics, however, comes with risks. The list of disadvantages reaches from duplicate work to inconsistent interpretation of data.

You can avoid these disadvantages by designing a central Data Analytics entity as part of your Data Office at an early stage, to create a common basis across all of these areas.

As you can imagine, such a design requires authority, as it would ask functional silos to give up part of their autonomy.

That is why it is worthwhile creating a story around this for your organization's Management Board. You'd describe the current setup, the behavior it fosters, and the consequences including their financial impact. Then you'd present a governance structure that would address the situation and make the organization "future-proof."

Typical aspects of such a proposal would be

- The role of IT as the entity with a monopoly for technology and with the obligation to consider the Analytics teams of the business functions as their customers

- The necessity for common data standards across all of those silos, including their responsibility within the Data Office

- Central coordination of data knowledge management, including training, sharing of experience, joint cross-silo expert groups, and projects

- Organization-wide, business-driven priorities in Data Analytics

- Collaboration bodies to bring all silos together on all management levels

## Data Science

You will find two fundamentally different Data Science setups in larger organizations. One is a centralized approach, with a number of highly specialized data scientists, often pulled together in one building or floor of the head office (and typically reporting into IT); the other one is a federated approach with data scientists across all business functions.

Obviously, both approaches have their pros and cons.

- The **centralized** approach allows for close collaboration between Data Scientists, and exchange about the newest tools or algorithms. But they often live in an ivory tower, detached from the business they are supposed to support.

- The **federated** approach, on the other hand, has Data Scientists exposed to the real business, including the questions that keep business folks awake at night. But they are separated from each other, with limited opportunities to ask questions, to exchange ideas, or to share solutions.

That is why I usually suggest a **hybrid approach**: A federated setup with Data Scientists reporting into "their" business functions, together with an intense data science network.

While most of the daily work takes place in the respective business functions, this model requires dedicated time fences for the data scientists to get together physically, from informal chats to cross-functional hackathons.

Such a network needs to be sufficiently strong to enforce visibility and transparency of Data Science treasures such as data sources, algorithms, and logic across all functions. This is to avoid black boxes created by single data scientists. After all, different data science teams may be functionally biased through their functional managers and driven by the task to work in favor of their business function.

To support the centralized aspect of a federated network, you might even consider introducing a matrix organization where all Data Scientists of the entire organization have a dotted reporting line into a Head of Data Science under the CDO.

Furthermore, it may be advisable to set up a small but strong central Data Science team, in addition to the "local" teams. This move can help create central expertise, allowing for the creation of a "Center of Excellence." Furthermore, it gives the Data Office direct access to Data Science without having to ask a business function to "help out." Such a team can deal with cross-functional data science, independently of any functionally determined priorities of the existing local teams.

# Data report management

Managers love reports – particularly reports with colorful charts and big fonts (and rightfully so). But to which extent do they really *need* them? How to determine the commercial benefit of a report?

If the management of reporting is governed centrally, this area can become more efficient.

Here, "governed" does not mean full central execution. But a central repository of reports and standards can help avoid duplicates and incorrect or inconsistent logic. Close collaboration with other central Data Management functions (such as Data Modeling or Master Data Management) helps increase the quality of reports.

Another objective of central reporting governance is cost awareness. Where reports are provided by IT on request of a business user, or where an existing reporting team creates them, these reports feel like "free of charge." They are not.

What would a centralized reporting team focus on? Here's my list:

## (i) **Cost-manage reports**

A typical price tag of a report would cover the cost of creation, cost of maintenance, and cost of complexity (which increases exponentially with the overall number of reports).

Even if cross-charging is not possible, full visibility of these costs will incentivize report consumers for retirement or sharing of reports.

## (ii) **Run reports**

Not all business functions want to set up their own reporting team. Some are too small or too focused on their core business objectives to do so.

This offers a chance for "voluntary" centralized reporting. You leave departments the choice whether to do it on their own or to ask the central team. As soon as the first two departments ask for centralized reporting, synergy effects kick in.

Midterm, a central reporting team will be able to offer standardized reports that should be cheaper for report requesters than tailored reports as they usually get developed by functional or geographical reporting teams.

(iii) **Develop and govern self-service**

Organizations benefit from an environment that allows for self-service reports that provide business analysts with the desired flexibility. However, self-service reporting also needs to be actively governed to avoid inconsistencies between reports created by different business functions.

You should ensure all relevant managers have access, including their mobile devices. And please work with IT to make it happen. This is a tangible thing, and managers will probably love you for this move.

(iv) **Foster collaboration and exchange ideas**

In decentralized reporting, the wheel gets reinvented many times. Be it the logic behind a report, a specific view on data, or even the usage of certain visualization tools – why wouldn't it make sense to share ideas and experience? A central team can facilitate this, thus shaping a "reporting community" within the organization. Whether you have your Data Consumer community cover this aspect, or whether you shape a dedicated reporting community is usually a matter of your organisation's size. If you have too many small communities, it will be difficult for them to reach the critical mass required for an active community life.

Finally, a word of caution: Centralized functions tend to become bureaucratic. This often goes hand in hand with degrading usability and slow response time, which impacts user acceptance more than anything else.

Monitoring such trends is, therefore, a key task of a CDO, be it through regular reviews with the central reporting team or be it through regular user surveys.

# Document Management
## A centralized approach can add value

Most organizations (not only the bigger ones) face a dilemma with Document Management.

Wherever a department has gone from paper to digital, they usually did it as part of introducing a functional software package, of which Document Management is a more or less integrated component. In each case, Document Management is mostly well integrated into the business processes supported by that software package – but usually entirely separate from all other business functions and processes.

Such an approach becomes a challenge where documents are of relevance beyond the single department. This is usually the case where end-to-end processes reach across functional borders. Think of a situation where Customer Service requires access to a customer contract, including the need to link this contract to a concrete case (e.g., a customer claim).

This example illustrates that there is no simple solution. You either integrate the processes and documents of each department individually, or you create an organization-wide Document Management system that is not fully integrated with functional software (the more so if the department has opted for a SaaS setup, i.e., its software is run by a third party outside the organization).

This challenge can certainly not be addressed by several business functions independently. It requires a joint effort of a central business entity (aka Data Office) and IT as the technical solution provider, in close collaboration with all business functions.

There are other structural challenges with digital documents that can definitely be addressed most efficiently through a central approach, for instance:

- Resistance: Human beings love the feeling of paper in their hands.

- Linking of documents and data across different business areas: Documents are available to all relevant functions. No sheet of paper needs to be imaged and stored more than once.

- OCR and automated tagging: Most documents contain all relevant information, both structured (such as account numbers) and unstructured (such as keywords in complaint letters from customers). A proper OCR process helps auto-tag documents. Systematic barcoding or 2D coding (e.g., of own forms handed out to customers) reduces the error rate of auto-classification.

- Technical synergies: One archive is cheaper to administer than multiple instances, even if the number of documents is higher.

# Documents in Data Management?

Let us define "Document Management" as the discipline of maintaining information in the form of individual pages that could be printed and read by humans.

Is Document Management a Data Office discipline?

Yes, it is, for two reasons:

(i) **Documents are data.**

Documents don't exist in isolation. They contain information that is meant to be interpreted. This even applies to drawings of abstract art which may not contain a single alphanumeric character.

Sometimes you cannot even draw a clear line between classical documents and structured data. Think of documents stored as a record of data plus a reference to a template (e.g. in the form of an XML file). Or think of documents that never get printed but are sent directly to a document repository, typically as a pdf file or an image.

(ii) **Documents don't exist in isolation.**

Documents don't make sense without metadata. You need metadata, such as identifiers, attributes, or tags, to find documents, to structure them, to archive or delete them (think GDPR), or to link them to other data.

# How to shape a Document Management team?

Documents are diverse in character, just as any other type of data is. They reach from digital receipts (in huge numbers but with a simple structure) to complex maintenance contracts (small number, text and diagrams, multiple signatures, various versions of the same document).

A Document Management team has two main tasks: Setting up Document Management solutions and managing documents on a daily basis.

(i) **Setting up a Document Management solution**

A Document Management team needs to understand the different requirements. They need to be able to analyze software solutions, both dedicated Document Management systems and document handling components of functional solutions.

Eventually, they need to decide whether one or more dedicated Document Management solutions should be targeted at or whether different functions should use the document components that are well integrated with their functional software solutions.

All of this requires a profound understanding of the business as well as good application knowledge. Document Managers need to answer questions like

- Does "one size fits all" work, or do we instead go with different solutions for different types of documents? Are there any synergies (joint data keys; or benefits from using a single workflow) that would justify a single solution?

- Can functional applications interface a central document repository, or should a central document hub point to documents in individual functional solutions?

- How do we organize central Metadata management in the case of distributed document repositories?

- In the case of distributed document repositories, how can we avoid duplicates of documents? If two different departments (e.g., Finance and Customs) require access to the same physical document, how can we ensure it is stored only once but accessible by both departments and referred to by both departmental applications?

## (ii) Managing documents on a daily basis

Many tasks around the handling of documents can be automated. However, there are limits.

While modern software can help extract Metadata, the last word often needs to be with human beings. And this is not only due to immature extraction logic. Sometimes the quality of metadata included in documents is poor.

Think of documents you have seen, which came with the Metadata field "author": How often does such a field disclose the true name of the author? In most cases, it contains the name of the person who created the document many years ago. This may have been a totally different document from which the one at hand has been copied and modified; sometimes it is even the "author" of the template, that is, the person responsible for templates within the Corporate Design department.

You can imagine that it does not make sense to automatically extract fields from existing Metadata in such cases. A person responsible for Document Management should monitor the situation. It is also necessary to work with Data Quality to set the right metrics and targets.

In addition, a Document Management function needs to be the interface to IT when it comes to document volume forecasts and performance requirements. Due to the mere size of image documents, a document repository is often the most storage-intensive database of the entire organization.

---

■ **Note** In smaller organizations, the entire "Document Management team" may consist of a single person. Document Management may even be one of several roles held by one person. However, the list of topics to be covered is relatively independent of the size of your organization.

---

# Data Quality

Data Quality has two organizational aspects.

One aspect is the setup of a central Data Quality entity with organization-wide responsibility. The other aspect is a general Data Quality responsibility across the entire organization.

## The central Data Quality team

A central DQ team needs to understand the business very well – as Data Quality is always linked directly to business deliverables.

And you would not want to have a bunch of academic "100 percent or nothing" people work here. In most cases, the target quality level is a function of costs and benefits, so that a target below 100 percent is often the better choice.

Luckily, you don't need to move your expensive and scarce data scientists to your DQ team. Data Analysts with a strong business background are definitely an adequate choice for a Data Quality role.

Here are some skills your DQ team will benefit from:

- A solid Six Sigma education (discussed later in this chapter) will help your DQ team determine root causes and to avoid shortcuts through "curing the symptoms."

- A good business understanding will allow the team to determine (and agree on) adequate Data Quality targets.

- Project management skills (Waterfall and Agile) help run their own Data Quality improvement initiatives and support business initiatives from a data perspective.

- Controlling-minded people will help you monitor, report, and follow up on Data Quality issues.

A significant risk that comes with a centralized DQ team is its perception as the "data police." Instead, you may want them to be perceived as the supporters of business.

Consequently, their focus should not be on ensuring people treat data properly but on promoting data as a valuable resource.

Ideally, the pressure to comply with all data standards should come from the consumers of the data themselves, not directly from a head office function such as the Head of Data Quality.

The key to achieving this is, next to creating a data-savvy organization, providing *transparency* of behavior.

The first step is to set Data Quality targets in collaboration with all business functions. Such a dialogue will see your Data Quality team in a moderator role, as the data users will ask the data creators to ensure Data Quality is good.

If, as a second step, you communicate the levels of achievement broadly, teams may try to improve their data handling just in order not to look bad.

## Data Quality across the organization

Everybody is responsible for Data Quality – but you can hardly ever reach everybody in the organization, let alone influence everybody in their daily routine.

That is where Data Stewards come into play. In addition to their regular work, they should feel like data ambassadors and promote Data Quality as a mindset.

## Organizing Masterdata Management

The management of Masterdata can be divided into design, coordination, and maintenance.

## Masterdata maintenance

Maintenance is a responsibility of a Data Steward, whose primary skill required is a good understanding of the functional background of the data.

You should never centralize such a role unless your organization is too small for qualified stewardship within the business functions.

## Masterdata design

Business functions know in principle what they need. But they are not necessarily able to translate those needs into a Masterdata concept. And they may primarily focus on their functional requirements.

That is why you need a centralized team to design an organization-wide, cross-functional target environment for Masterdata as well as the path toward it.

The entire design work, including the selection of the best possible style(s), should be done in close collaboration with Data Architecture.

You will not always enjoy the luxury of being able to redesign your Masterdata environment from scratch. In most cases, you will have to live with legacy environments, with disconnected data repositories, with the technical need to maintain the same data in different places, and so on.

Finding a setup for Masterdata to run smoothly under these circumstances is a big challenge – possibly the biggest challenge for a central Masterdata Management team. But you should by no means wait until a perfect environment is in place. This may take years even in agile organizations, due to the enormous number of technical dependencies found in most organizations.

My pragmatic recommendation is to set up a team that focuses at 75 percent on running and optimizing Masterdata under the given constraints. The remaining 25 percent should be used to design the target environment, based on business requirements, technical opportunities, and findings from current operations.

## Masterdata coordination

Coordination is a critical responsibility of a central Masterdata team. An uncoordinated situation will not allow for efficiency and consistency for the following reasons:

- Each area of Masterdata is often owned by a single function – but the data is usually required by multiple functions.

- Furthermore, while maintenance responsibility for each Masterdata domain should be with one function only, this is not common practice in many organizations.

- Standardized measurement of Masterdata Quality and the determination of the minimum quality level must be determined across all business stakeholders of Masterdata.

- Each organization needs a single point of contact (SPOC) for everybody to ask for the availability of Masterdata – be it for the operation of machines or be it for process automation (Robotic Process Automation, RPA).

- Alignment, balancing, and prioritization of the needs of various business functions should be done by a party that has no vested functional interest to avoid a biased result. The same applies to the development of decision proposals.

# Data Project Office
## Yet another overhead function?

Data-driven organizations run data initiatives. As with other data-related activities, this should not be done in functional, organizational, or geographical silos. You'd instead want to see organization-wide coordination, prioritization, and collaboration.

In a well-managed organization, there is a clear commitment to having a single Data Supply Chain, shared by all parts of the organization. That is why nobody should be in a position to change it unilaterally, just as no part should be allowed to build a parallel Data Supply Chain.

You will hardly ever find a data project with an impact on a single business function only. Most data projects are of cross-functional relevance, just as data is cross-functional by nature. And even different dedicated data projects usually have strong interdependencies.

As a consequence, organizations need to manage all changes to their single data supply chain to achieve consistency, including a consistent CDM.

Such changes happen through projects, justifying a dedicated Data Project Office as part of the Data Office.

# Responsibilities of a Data Project Office

Typical Project Office activities are, of course, applicable to a Data Project Office as well.

As a general principle, a Data Project Office would take care of administrative tasks which would otherwise burden the individual initiatives. Planning, coordination, and communication are additional, well-known tasks.

But a Data Project Office would also take care of the following topics:

- It would ensure everybody follows Data Standards and Principles, and it would help check for compliance.

- It would monitor not only the execution of activities but also subsequent delivery of forecasted functionality and benefits. This task is often neglected when dealing with data projects, considering the difficulties in quantifying data-related benefits.

- Finally, it would support the transition from a possibly outdated, traditional project culture toward a modern project culture with a desire to consciously balance agility and sustainability.

# Focus areas

More than any other team, a Data Project Office has to be set up with a strong focus on the following two aspects:

(i) **Agility**

In today's project management culture, hardly anybody expects all requirements to be gathered up front. More and more frequently, projects are expected to allow for business requirements to be determined and prioritized during development, including their immediate implementation and deployment.

While agility is an essential component of any modern project management, it requires particular attention when dealing with data: Agility with a primary focus on speed will jeopardize interoperability and foster ambiguity issues. This is why agile data handling requires discipline – in managing technical change and in documenting it.

You may want to have a look at Chapter 20 for some thoughts on DataOps, which can be an effective methodology in ensuring well-managed agility.

(ii) **Collaboration**

From a data perspective, no workstream should work in isolation. Interdependencies are found both within any single waterfall project and between different phases or sprints of an agile project.

On closer examination, most seemingly independent projects turn out to have data interdependencies.

# The Data Project Office within the organization

Any Data Project Office would be responsible for pure data projects as well as for data aspects of other projects, especially in organizations with no other project management organization in place.

In organizations with an existing, cross-functional Project Office, a Data Project Office still has its justification. Here, however, it would play more of a **supportive** role: It would cover all data-specific project management activities that a generic project office would not have the know-how to handle.

In case of a coexistence between a general project office and a Data Project Office, it is vital to agree on a well-defined split of responsibilities.

This would already be done during the setup phase, and it would include a commitment to collaborate closely.

# Setting up a Data Project Office

All responsibilities described in Chapter 9's section "Data solutions and projects" would find their home in this team.

The right people for such responsibilities have data as a secondary strength only. They primarily need strong project management skills (both waterfall and Agile), so that they can run their own projects **and** support business or IT projects from a data perspective.

You may wish to have a trustworthy data representative on any project or scrum team – not necessarily as a full-time resource but with access to all details. They must be able to play by the rules of those teams while enforcing data requirements in a nonconfrontational way.

During its introductory phase, the Data Project Office should focus on its role as a supporter to gain acceptance.

Enforcement of compliance should be increased gradually, as people begin to understand the value of this new entity.

# Data service function

It is good to provide the organization with a single point of contact in data matters.

You are serving the organization. As a consequence, why not respond with a backline and frontline organization, as in professional Customer Service entities?

As usual, different roles do not necessarily mean different persons. Depending on the size of your organization, the entire Data Service function may consist of one single person. (You may wish to ensure, though, that there is a backup for absence related to holiday and illness.)

The key purpose of the entire setup is the provision of contacts for people to talk to, and of documentation for people to read.

But you also need back-end functions which follow up with the experts, get back to the requestors, and finally connect those two.

The front-end function is the first point of contact. It also manages any service-level agreement (SLA).

A workflow tool helps follow up and avoids orphaned requests.

In big organizations, it may make sense to log the effort and to avoid the occupation of critical resources through less relevant requests.

## Business helpdesk

In your organization, do all employees with a data-related question (or proposal, or complaint, etc.) know whom to talk to?

Do they need to know?

Ideally, there is one single point of contact for all types of data-related topics. The Data Office should be in a better position to classify cases than any requestor.

That is why it makes sense to set up a business helpdesk for data topics, similarly to the helpdesks IT organizations have been using for decades. (You might even consider a joint helpdesk with IT to make it even easier for requestors.)

Who would work on such a helpdesk?

No helpdesk agent can answer all content-related questions – at least not in a first call.

That is why you need generalists who understand all content-related questions sufficiently well to be able to determine the right expert.

And how would you position your data helpdesk within the organization?

Even if you say your helpdesk is meant for "everything data," some people will not understand what it entails, and you might need to help them understand the scope of such a helpdesk.

That is why it is a good idea to involve business people in the development of your helpdesk, including the design of processes and (ideally) a supporting workflow solution.

It is beneficial in this context to communicate proactively, offer training, and provide examples. No data question should ever remain unasked because someone did not dare ask!

For the Data Office to gain credibility, a data helpdesk should pick up reported topics even if they are not formally "in scope."

In fact, you cannot blame anybody for not knowing what falls under "data." It is a gray area, and while several topics fall under "data" in some organizations, they are with different functions in other organizations. Typical examples are "data-related fraud" or "data security."

Sometimes there hasn't been anybody in charge of a new topic within your organization. You might use the opportunity to clarify responsibilities, ideally through the Data Governance bodies, that is, the Data Management Council or the Data Executive Council.

# Data organization contact

Data handling comes with a lot of noncontent-related aspects. They reach from documentation to the Data Office Intranet site, including the adequate provision of online information.

An essential aspect of a Data Office's daily work is the follow-up on Data Concessions and Data Quality issues. Who was responsible for which action, and when was that action due?

You might gain efficiency by bundling all of these activities in a dedicated entity. This team does not need to consist of data specialists.

As with other areas of your Data Office, the optimal sizing depends on the size of your organization. A larger organization will benefit from a dedicated Project Office for data-related activities, including project management resources. In smaller organizations, the CDO's assistant could be the contact person for any organizational data topics. Everything in between is possible as well.

# Attracting and retaining experts

Data experts are hard to attract and hard to keep on board. This is particularly valid for (but not limited to) data scientists.

There are many reasons for this situation. Of course, data scientists are rare. I mean, real data scientists, not rebranded analysts or highly gifted spreadsheet jugglers. I am thinking of those with a mathematical background, programming skills, creativity, curiosity, and discipline.

But how about those few who exist? Let's have a look at ten aspects that influence an expert's decision to join and stay or to leave.

I am focusing on Data Scientists as this is the most challenging group of employees. Most aspects can be applied to other data experts, though.

## Diversity is beneficial — as an outcome

I have had people from more than 30 nationalities working with me during the past 20 years, coming from four different continents. And, on average, 50 percent of my direct reports have been female.

Did it pay off? Yes, definitely! Whatever the topic was, I have always had more than one perspective. People with different backgrounds look at things differently, and they often come to surprising solutions. Discussions within a diverse group are generally more fruitful.

But how did I get there? Did I set gender equality targets? Did I work toward a balanced representation of different nationalities or ethnic groups in my teams?

I guess I am pretty demanding when selecting employees for my teams. I want them to have excellent knowledge and intellectual potential, the right attitude, and the necessary soft skills. Gender, origin, and color of skin, however, have never been any of my criteria.

So, why did my approach always result in very diverse teams?

It's because the reality is diverse, and no gender or nationality is superior to any other. In addition to the "hard" criteria I mentioned earlier, I always search for characteristics complementing those prevalent in my team. This approach resulted in a broad diversity of gender, origin, age, and other attributes that are often subject to discrimination.

If, instead, you set hard equality targets, you will often be forced to go for the second-best candidate. This is neither good for your organization, nor is it fair to the best candidate. Let us be honest: Rejecting a person just because it is, for instance, a white male in his 50s is discrimination as well.

That is why I don't think it is advisable to fight inequality by practicing "reverse inequality" – even if your intention is laudable.

To summarize my experience: The power of diversity unfolds if you explicitly allow for it, not if you enforce it! This helps you achieve both diversity **and** the best possible teams.

# Everybody wants to join Google

Is your organization sexy? I mean, *really* sexy?

Let's be honest, most organizations are not attractive to outsiders, no matter how good employee motivation and culture may be inside the organization.

As with all other people seeking a job, Data Scientists rate potential employers. Of course, on top of a Data Scientist's priority list is data science. But what comes next in parallel to the working conditions and content?

It is "Does it look good on my CV?"

A life-long loyalty to an organization is not on top of a Data Scientist's list. Many of them already think of their "possible next employer."

Can you imagine a Data Scientist proudly stating "I am doing data science for a mid-size company that produces injection pumps for asphalt milling machines!" Compare this with "I am part of Amazon's machine learning team!"

To attract good Data Scientists when you are neither Google nor Amazon, you would advertise the opportunities of the role itself, not how successful your organization is. Success, after all, is a matter of perspective.

# The sweeter challenge next door

The grass is always greener on the other side of the fence?

It probably isn't, but how should somebody know who cannot compare?

The key to retention is satisfaction. Happy employees compare their current position less with theoretical alternatives.

Conclusion: A happy Data Scientist will stay.

Wrong.

You may have heard yourself state in strategy meetings that "Great is the enemy of Good." Well, did you know that, particularly for Data Scientists, **great jobs are the enemy of good jobs**?

As a consequence, you need a retention strategy with ongoing observation and execution. The "best job on earth today" may become the second-best tomorrow.

First of all, you would aim at making your Data Scientists happy.

Secondly, you would also explain why they can be satisfied and what is so specific to working for their current organization.

Such reasons may be specific to the organization, or they may apply to the industry you are in. As a positive example, look at the following statement from Angeli Möller, a leading Data Scientist at Bayer Pharmaceuticals: "Driving data science in healthcare is so rewarding because of the significant benefits to patients" (Möller, 2019).

It is generally recommended to pull Data Scientists out of their comfort zone. Show them your organization's operations. Let them listen to your customers. Explain to them what the data means they are working with.

## No meetings, please

"A team meeting is a team meeting. Everybody has to attend!" You may have heard that statement. Maybe it was even you who said it.

To many Data Scientists, meetings are a total waste of time – even more than to other groups of employees.

It often helps if, whenever possible, you make their participation in meetings – big ones or small ones – *optional*.

In this case, you will have to introduce alternative means of communication to reach your data scientists. These means may include individual verbal updates at their workplace. Yes, it is more work, but Data Scientists usually appreciate this effort.

It is not a bad idea to prepare relevant information for Data Scientists similarly to how you prepare updates for the Management Board – brief and to the point.

# Where's the infrastructure?

An excellent Formula 1 driver will be reluctant to join a team that does not have a competitive car.

Similarly, an excellent Data Scientist will expect all the infrastructure to be available, ready to be used.

Data Scientists who would love to build a Data Science infrastructure from scratch are different from Data Scientists who want to do data science. I'd assume there are more of the second kind.

If you really don't have anything in place that you can base a Data Science team on, try to find those few that would love to be pioneers. Such Data Scientists might be a bit more expensive (they have to be experienced in people management, and in technology *and* data science!) – but you can expect them to become your future team leads.

They may also become a bit more loyal than the average Data Scientist – perhaps not to the organization but to the data science ecosystem they establish.

# Playground vs. strategy

A beautiful and well-equipped playground is sufficient to attract little children. But does this approach work for Data Scientists as well?

For most of them, it doesn't.

Data Scientists usually want to work toward a target – the bigger picture if you will. Getting trained and learning new things are certainly important to them. But most of them want to do all of this for a reason.

That is why you should share your organization's business targets and its data targets with your Data Scientists.

But not through a meeting…

# Detached from business

You sit in the data science corner, play around with data sources you can get hold of, and wait for business folks to come with requests?

Some Data Scientists may indeed like such a setup. Most of them, however, would like to be part of a bigger picture.

Furthermore, Data Scientists are usually most productive in interdisciplinary teams. After all, they should be able to address the true challenges of their business colleagues.

That is why I always recommend that most Data Scientists sit close to their business counterparts – maybe even become member of a business team.

At the same time, it is vital to have all the Data Scientists of your organization feel part of one Data Science community.

## Little, stupid jobs

New Data Science teams need to fight for acceptance within the business community – before being asked for the big, expensive projects. That means starting with small steps. In reality, this happens through little favors: "Please find me X," "Are A and B correlated?", "Can you remove the outliers from this list?".

A data scientist who has just graduated over "convergence criteria with multidimensional gradients" must feel offended here. Why have they learned such fancy stuff, just to go back to the pre-university level?

Please try to influence the team's motivation. It goes back to the basics: My primary target is to make my customers happy.

## Even Data Science can be boring

Imagine a Data Scientist who has finally found the perfect place within an organization, serving a bigger picture through full alignment with the business teams. All good?

Not necessarily.

Data Scientists may find themselves in a "user" position. They have learned all the algorithms and logic, and now it takes one line of code to initialize a model in TensorFlow, and another line of code to train it, using data that was gathered by other folks.

Think of a typical developer of computer games: They usually don't enjoy playing the games they developed. Excellent Data Scientists often feel the same.

This is where hackathons come into play and where you might set up data science competitions with Data Scientist teams from other organizations (not your competitors, of course). Under such competitive circumstances, it may even be fun to search for useful data on the Web and to try to understand it properly.

# Recognition?

How can a Data Scientist be recognized for all the incredible brainwork hardly anybody else could do?

I mean, how could they *truly* be recognized? I am not talking about the kind of awe mixed with aversion and anxiety people face after admitting to having studied Physics or Mathematics!

Tell the stories! Communicate success stories to your business folks, in business language. Tell people that this is something computers and algorithms alone couldn't do.

# Data Scientist vs. DB Admin

Is everything around processing data a Data Scientist task? At least everything that a "normal" member of a business team cannot do?

If a Data Scientist shows competent in all aspects of data, they will quickly become the first person to talk to in all data matters.

Unfortunately, this comprises data activities for which a Data Scientist is overqualified or work that usually requires entirely different qualifications.

A lot of these activities would fall under IT. That is why data handling should be a focus area of the initial alignment tasks between the Data Office and IT. And, whatever the agreements are, they should be documented and communicated beyond these two teams. It also helps if the Data Scientists know whom to refer to if asked for support outside the Data Science domain.

# Curing the world's hunger

But what happens if, within an organization, a Data Scientist comes to fame for delivering great results?

Expectations will go up. Particularly, business folks who, for the very first time, see how a neural network "learns" pattern just from (seemingly) chaotic data are often deeply impressed.

How would you address unrealistic expectations?

Manage those expectations as part of your internal communications. The entire staff needs to understand

- What Data Scientists are supposed to do (and what you should *not* expect them to do).

- What data science can achieve, and what is not possible. (Everybody knows that nobody can determine next week's lottery number. Some requests come close, though.)

You might not want to demystify Data Science entirely. Your experts should receive the respect they deserve. But your business colleagues should certainly be prevented from seeing Data Scientists as the first successful alchemists in history.

# Six Sigma

You might hear people say that Six Sigma is outdated or that it has been replaced by Agile.

Don't believe them. No methodology has shown superior to Six Sigma to date. And Agile is a different ball game. Agile and Six Sigma can work together very well.

Let's have a look at how Six Sigma can help improve data handling.

## Six Sigma and data

People tend to cure the symptoms instead of fixing the root causes. This is partially due to the wrong motivation – there is an expectation to fix problems quickly. But another reason is that people think they know the cause. It looks so evident at first glance.

But what people consider the cause of a problem is in reality often a symptom on its own, caused by a problem that runs deeper.

Many organizations who understand this situation introduce Six Sigma as a core principle and methodology across all business functions. This comes with both a set of tools and a different way of thinking.

But what does Six Sigma have to do with data?

The connection is twofold:

- First of all, you can hardly run any root cause analysis without proper data.

- Secondly, the Six Sigma methodology is well suited to work on data-related problems as well.

Interestingly, most data-related problems lead to root causes beyond data. Typical root causes for data issues reach from lack of process knowledge to inadequate incentivization leading to the wrong behavior.

# Setup of Six Sigma within the Data Office

If there is a Data Quality team within the Data Office, this is the place to run Six Sigma initiatives on data issues. If there is no such team, this is an excellent reason to set it up.

A Data Quality team would always act as a custodian of Data Quality, not as an owner. It is the business leaders who have to take ownership. The business owner of a problem should not be able to "outsource" the analysis and the necessary fixes to the Data Office.

This makes clear again why it is so important that the Data Office has the authority to

- Run Six Sigma initiatives

- Request action on determined root causes

- Assemble cross-functional teams

- Accompany the projects by measuring and consulting

without the consent of all impacted functions.

Why? Imagine the following situation: Team A introduces a new process that reduces the workload of team A. At the same time, this process causes a problem for team B that works with the output of team A. The impact on team B may be more significant than the savings for team A.

In other words, the new process is a bad idea from a shareholder perspective. If a root cause analysis starts with team B, it will quickly discover the situation at team A. Team A should obviously not have the authority to say "Stay out – this is none of your business!"

This example makes it clear that such a Six Sigma team must by no means sit within any of the business functions, to avoid such an unhealthy bias. The Data Office, if set up correctly, is a perfect place as it reports into the functionally independent Chief Data Officer.

As soon as you have the required authority, the necessary Data Quality team structure, and the right Six Sigma experts on board, you can define the Six Sigma process for data-related issues.

# A typical DMAIC data process

Within Six Sigma, *DMAIC* stands for the five phases of a classical improvement process: **D**efine-**M**easure-**A**nalyze-**I**mprove-**C**ontrol.

These phases can easily be used to shape a **data-driven Six Sigma process**, as outlined in the following. You will be able to see from these steps how critical a strong mandate and a stable governance structure are.

## Define

The process starts with the registration of a problem.[2] Processes should be prepared for the following sources:

(i) **Issues reported through the data network**

Problems coming up in a single country may be considered too minor or too expensive to fix. If we see that they occur in multiple places around the world, we may have a good business case to act.

(ii) **Issues indicated through DQIs**

The more we will measure Data Quality, the more we will make weak spots visible. A proper root cause analysis will help avoid premature action.

(iii) **Systematic process assessment**

It is always a good idea to have all relevant processes reviewed systematically, based on thorough discussions with Global Process Owners and their teams.

(iv) **Previously known issues**

Sometimes issues have been known for a long time. They might have been gathered or reported in the past but not acted upon. We now have the chance to assess them and to make them visible. If the case is good, it will be difficult to ignore.

---

[2]Here you will need to define who is authorized to report a problem and to whom it has to be reported. It is important to ensure that nobody can prevent the report of a serious problem at this stage.

### (v) **Top management requests**

Sometimes requests come straight from the top, for example, executive complaints received from key account customers or an executive's disappointment about the unavailability of data.

Once registered, the issue will be allocated to a Data Quality Improvement Analyst (DQIA) within the Data Office, ideally a member of the Data Quality team.

The DQIA will (pre)assess the case: Is it really a problem? Does it require an analysis? Is it part of a previously reported problem? Is this an opportunity to improve customer experience or business performance?

## Measure

The status quo gets measured using adequate metrics (Data Quality Indicators/ DQIs).

The quantification of the problem is also part of this phase – it is a precondition for the ability to measure progress later.

The impact gets quantified by the DQIA, and a first business case is built: Is the effect of the problem sufficiently severe to start a systematic analysis?

If the impact justifies the continuation of the initiative, the DQIA takes the next step.

## Analyze

A root cause analysis is executed by the DQIA, in collaboration with impacted people from all relevant functions. The outcome is a project recommendation, including possible workstreams and their functional owners.

If necessary, a data decision body is asked to endorse a preproject first, aiming at creating a business case, determining IT costs, project duration, and resource requirements.

The data decision body will finally be asked to prioritize the case, and it will provide a go/no-go decision.

## Improve

After the final approval and funding, the actual execution of the improvement project starts.

Project types can be as diverse as

(i) **Validation against the glossary**

Such a project can help resolve misunderstandings. It usually starts with stocktaking of existing documents and IT applications, followed by a harmonization or standardization project.

(ii) **Data Quality assessment projects**

Such a project starts with the translation of a reported business problem into a data question.

Close collaboration with IT is required to ensure adequate data management solutions are used to locate and quantify the issue.

The output of such a project can be the institutionalization of the assessment (e.g. through Data Quality dashboards) or the proposal for another project to address the determined issues.

(iii) **Data Quality improvement projects**

Such projects often follow Data Quality assessment projects – which ideally provide cost, resources, duration, and success criteria for such an improvement project.

(iv) **Engagement and knowledge initiatives**

An assessment often reveals a lack of knowledge or awareness within the workforce. You can address this through a dedicated project aiming at training, improved information exchange, or employee engagement.

This is a beautiful example of the fact that Data Management is not a purely technical discipline.

Note that none of these projects does necessarily need to be run by a member of the Data Office. Each project should be run by the best-suited business function – which is either the most impacted one or the one in charge of the proposed activities.

Different functions can run such subworkstreams – but one of them must be in the lead and provide the project manager. The contact person within the Data Office will follow up on the status and progress.

## Control

After completing the improvement project, the result needs to be measured. This phase will be run by the Data Quality Owner of the initial problem. That owner will report these results to the body that has approved the project. This body decides about project closure, project stop, or a restart at an earlier phase (e.g., to refine the analysis or to propose a different project setup).

Where applicable, the changes implemented by the project will be communicated to a broader audience. Policies and procedures may need to get updated, and training may need to be modified.

# The Psychology of Data Management

# Typical Challenges of a CDO

*"You may well have data, Smithers,*
*but I have strong opinions,*
*and I pay your wages"*

**Figure 12-1.** What is the standing of data in your organization?

© Martin Treder 2020
M. Treder, *The Chief Data Officer Management Handbook*,
https://doi.org/10.1007/978-1-4842-6115-6_12

# Why is it so hard to be a CDO?

*The business case was great, the story compelling. The superiority of the data solution compared to the conventional approach was proven. Funding was granted, the project started.*

*Preparation went on and on. IT complained about lack of ability to test. First results were obtained but couldn't be validated. Data experts were either busy asking for support or sitting idle. Sponsors remained silent. The project ran out of time, then out of funding. It stopped silently, and the already implemented functionality remained unused.*

Does this situation sound familiar? Welcome to the club!

It is not lack of technology, great ideas, or a good strategy that make a CDO's work so challenging. It is situations like this.

But what has actually happened?

In this case, it was the frequently observed mixture of the business team feeling insecure and their manager feeling threatened: "They want to tell us that computers can do better than us, despite our decades of business experience...?"

As a consequence, the team didn't provide the right information, no accurate data, no feedback, no proper testing, and so on.

None of them was afraid of personal consequences – the project didn't have an engaged executive sponsor anyway.

**Not the technology or lack of business value made this project fail – the nature of human beings did!**

Is this story the rule, or is it a tragical exception?

Unfortunately, you almost have to expect situations like this in any organization without data being deeply rooted in its DNA, even if it has just introduced a CDO role and committed to funding a Data Office.

To understand the reasons for failure, we should first have a look at the motives of an organization that intends to introduce a Chief Data Officer.

**So, how is a CDO born?**

This is where the challenge starts: Hardly any organization introduces a CDO because the Board understands (a) what data can achieve and (b) what needs to be done to get there.

Instead, big corporations often decide to introduce the role of a CDO as it seems "the right thing to do these days". Often, one of the Board members sees the need to act after hearing or reading somewhere that "data is key to success" or "data is the oil of the twenty-first century."

Let's be clear: Such an organization is already light-years ahead of other organizations that still assume that their twentieth-century setup is sufficient to deal with the ever-growing amount of data and its opportunities.

However, even in many organizations that created a CDO position, the top management does not know what "data" really means and what precisely a CDO should be responsible for.

This often leads to vague expectations toward a CDO, such as "Please transform us into a data-centric enterprise! (And let us know once you are done.)"

### Why do CDOs often leave after a few months?

(i) **No job description**

It is difficult to be rewarded for a good job if nobody in the enterprise knows what "good job" really means. And often great improvement in data handling remains invisible as all the tangible benefits are claimed by functions that benefit from the Data Office's achievements.

(ii) **No established role within the organization**

Several Board members may welcome the CDO with open arms – many of the other leaders may not. They don't see the gap the CDO is supposed to fill, and they may even perceive this new position as internal competition.

(iii) **Irrational behavior**

What seems common sense to a Chief Data Officer is not necessarily the logical conclusion of each single Board member.

Alan Duncan, Research VP for Data and Analytics Strategy at Gartner Group, summarized the first learning of almost every new CDO to be "Decision-makers are emotional, not rational. Having the answer is not enough."

(iv) **But high expectations!**

You might not be expected to cure the world's hunger. But expectations often come close. Your organization wants to address the challenges of the digital age, and top management expects the CDO to achieve this target miraculously.

These points illustrate why a good CDO is not only expected to be a knowledge leader. You will read in the following chapters which other leadership qualities I have found to be essential to become a successful CDO.

To set the scene, I am beginning with a list of challenges I have seen many CDOs and other data leaders face.

You might also have come across one or more of the following situations:

- IT folks working on data are mostly not business oriented, not proactively determining the next necessary step.

- Most data experts are spread across all business functions.

- Data access to finance data is restricted on purpose.

- You face tremendous difficulties in getting your team to communicate what you are doing all day.

- IT still fulfill wishes without asking for the underlying business problem.

- InfoSec folks are not working together with data folks in business departments. They are taking care of information security on their own, the way they have always dealt with it.

- A Data Management team that is part of one business function experiences good collaboration with that department, but the data folks feel systematically being kept out of other business departments, with their work being undermined/duplicated by those departments.

These are some of the symptoms of an immature data culture within an organization. Established mechanisms to deal with such situations are missing.

If you don't feel totally out of place, please read on.

In Figure 12-2, I have grouped my observations into eight different categories. Each of them may block your journey toward a data-driven organization, just as big rocks can prevent a river from reaching the open sea.

**Figure 12-2.** The CDO meandering through corporate challenges

How do you address these challenges? In many cases, there is no straight path, so that you might have to meander, like a river in hilly territory.

Here are my proposals, taken from the real-life experience of CDOs and other data leaders.

# Struggle for supremacy

Your organization has developed its mode of operation in the past. It often has taken decades to devise a mature set of policies that ensures all topics are covered.

Now a Chief Data Officer enters the scene, with unclear responsibilities. Typical responsibilities of a CDO have so far been with IT, with Legal, with Risk Management, with Finance, and so on. Nobody wants to lose influence or authority to the new Chief Data Officer.

If this conflict does not get solved, you will observe a never-ending sequence of daily battles for supremacy. It becomes difficult if not impossible for the Data Office to work effectively as a data team's success depends on collaboration. How to address these challenges? Here are my thoughts:

(i) **The mandate.**

My first recommendation may not sound complicated, but it is challenging in many cases:

Ask for a clear mandate from the top

- Very early during the process of introducing a Data Office

- As nonnegotiable and untouchable as the mandate of, say, the CFO

- Communicated by the Board itself (instead of letting you run around holding your mandate up like a license)

It is essential for you to explain the need for this approach to the Board. And you might want to suggest that you take care of the wording of the mandate yourself. After all, you are the expert.

Don't forget to offer a few options (all of which you would consider acceptable) to demonstrate that you accept the Board's authority and that they have the last word.

This approach is also an early litmus test for the readiness of your organization: If the Board refuses to give you this kind of initial support, you should start responding to the headhunters again!

(ii) **Each topic already has a historic "owner."**

But the mandate alone may not be sufficient. You will have to deal with those that perceive you as a competitor.

I recommend two general rules for this situation:

No battle for single areas. Don't make enemies. You will need allies.

Divide and conquer! Give up noncritical areas.

The need for further activities depends on the concrete situation. You may face the following two variants of the challenge:

**Variant 1: "Stay out! This is my data!"**

Here, you might not need to escalate or ask for board-level support.

After all, someone is preventing a portion of the organization's data from becoming subject to professional data handling.

Make the situation transparent and let others complain: Those who don't get the proper data and excellent data services others get.

Variant 2: "I don't report into you!"

Do you primarily want to enhance your power base? This is generally not a helpful attitude of a CDO. You have a problem now, and rightfully so!

Otherwise, use your credibility to build trust and long-term relationships. Start with those of which you have the impression that they have the strongest positions within the leadership network. And be honest! After all, you want to support the other functions, not to become their boss.

# Lack of awareness

If you hear a fire engine coming, but you don't see any fire, you would wonder what they are about to do. You have two options now:

- You ask them to go away because you don't see a reason for them to be around.

- You wonder whether there is a reason for their presence that you are not (yet) aware of.

After a short moment of contemplation, you would probably agree that the second option is a more adequate one. Maybe the reason for the fire engine to move out is not as apparent as an open fire.

The first arrival of a CDO at an organization is often a comparable event. And you will observe the same two reaction patterns: Some will assume there are good reasons for a CDO to come, while others will consider a CDO nonsense as they don't see a burning fire.

You might want to know which of the relevant stakeholders belong to the second category. Typical statements of such people, often made implicitly, are

- "Which problem do you intend to solve?"

- "We have been very successful without a CDO for the past decades."

- "You are creating duplication as everything you consider doing here has already been taken care of."

Let's assume these people don't know better, that is, this is really what they think.

A good reaction in such a situation is to explain the background in a very modest way.

You can invite groups or individuals (the latter is recommended for top-level executives) to an exchange about the topic. Please do not pontificate – have an open dialogue instead. Listening and asking questions are the main elements of such a conversation.

As a result, you might not only be able to tailor your story better to the situation of your counterpart – you might also learn to better understand the organization, including its unwritten laws.

Content-wise, a lack of awareness manifests itself in two fundamental questions: "WHY?" and "Why NOW?" How do you respond?

## WHY should we manage data?

The ideal response to this question is a fact-based description of the *overall* rationale – your main story!

This story should not be about technical, data-centric arguments. "We need clean data!" doesn't mean too much to most executives. Instead, you may wish to provide a commercial rationale, along the lines of:

The primary objective of the Data Office is to help the organization

- Earn money
- Save money
- Stay ahead of the competition

In other words, the CDO intends to make everybody else within the organization more successful.

As far as possible, you'd describe both the gaps in the current way of working and any missed opportunities within the organization. You also say why you have come to that judgment. (To be credible here, you should first talk to the people at the front line before scheduling meetings with executives!)

The better you manage to determine the business challenges of those leaders and to focus on developing solutions to those challenges, the higher the chances are for you to become their accepted problem solver.

## Why should we manage data NOW?

In response to this question, you might wholeheartedly agree that there were times where active data management was simply not necessary: Which organization would have required a Chief Data Officer in the 1990s?

However, "The Times They Are A-Changin'," as Bob Dylan taught us decades ago.

So, what has changed?

First of all, people know that the amount of data has grown exponentially during the past couple of years. It is also clear to everybody that this development is not going to stop any time soon.

Secondly, customer expectations have changed. The eCommerce boom has taught many customers what is technically possible. Why should they settle for less?

Lastly, business decisions should no longer be based on gut feeling alone – no matter how experienced the owner of the guts may be. The world has become too complex for this management style. Furthermore, the ability to determine the best decisions through data is increasing as rapidly as both the volume of data and the quality of supportive algorithms do.

# Business silos

Professional data management is cross-functional by nature. Ambiguity and duplication can be avoided.

On the other side, functional leads may not see the value of putting energy into something primarily other departments would benefit from. They are measured strictly against their own teams' targets, and they wouldn't want any energy to be diverted from reaching those targets.

This expresses in various ways. The two variants I have come across very often are as follows.

# Variant 1: "We know best what's good for us."

This may, in fact, be true! Not everything may need to get centralized.

Such a statement, if made frequently, should provide you with an excellent basis to jointly work on a governance model with strong local or functional teams.

Think of central enablement of strong functional teams, sitting close to their business counterparts.

Think of achieving synergies by linking people through networks.

By giving in to people's request for autonomy in certain areas, you should seek for their agreement to centralize other topics, for example, in the field of standardized verbiage and processes.

# Variant 2: "I am faster if I do not need to align with others."

Perhaps you have made good progress in ensuring functions don't bypass you when asking for IT support. However, they often don't even need IT support – because they have already built up their own "Shadow IT" groups or structures.

But how to respond?

It is true: Alignment takes time. But what are the consequences of missing alignment?

- Activities and communication are functionally limited.
- Opportunities to exchange with other functions are missed.
- Different departments come to different outcomes – for the same topic.
- The origin of data may be dubious. Data may be misinterpreted.
- Work is being done that has already been done before.

However, it may not be sufficient to come solely with fact-based arguments in such a case. Nonalignment remains faster and thus attractive.

Two additional activities may help here. One is a solid business case that shows the long-term benefit of sustainable data work.

The other one is the creation of a win-win situation: You would first want to find out *why* a functional lead or a project manager wants to speed up. In many cases, it is the pressure to keep committed timelines. Such a situation allows you to team up with the project manager: You can jointly get it right the first time while getting the timelines amended to make it possible. After all, there are good reasons for an amended timeline, and they are clearly not caused by bad project management!

This approach helps you achieve a few things. The project manager gets enough time, together with the possibility to deliver high-quality work. The vast majority of project managers prefer the latter if you give them a true choice.

Furthermore, if successful, you might have won an ally, for similar cases in future. And you are given an opportunity to demonstrate that future projects become easier, faster, and cheaper due to the solid foundation that this first project has laid.

# Lack of ownership

Data ownership is one of the concepts in data management that not many people are familiar with.

But even ownership in the sense of "Let me take care of it!" is difficult to achieve between business teams. As soon as a business concept is perceived as "technical," people are often happy to outsource the responsibility to IT.

And, to be honest, why should anybody do a job that somebody else does voluntarily? In most organizations, IT has implicitly accepted data ownership during the past decades, as and when it became necessary.

Shouldn't you just leave it like this? Well, if you intend to leverage the full power of data, you cannot leave decisions about the business model and its data representation to experts in technical data modeling, software development or database design.

In fact, "Ownership starts at the top of the organisation," as Ole Busk Poulsen, Head of Data Governance & Information Architecture at Nordea, stressed at a data conference in Vienna in 2019.

But how to make people assume ownership? Again, the key is the creation of win-win situations. This implies a strong Data Office. You'd work with the business functions to help them understand how to translate their business concepts into a data structure.

At the same time, you'd explain to IT that you are making life easier for them: Instead of having to talk to various business functions in parallel and to balance conflicting demands, they would have the Data Office as their single point of contact and a facilitator between different business functions. This allows them to concentrate on their core responsibility: Providing first-class IT solutions that meet the business demands.

Figure 12-3 illustrates the win-win situation for a Data Management organization.

- **Easier for IT**: A single contact for all business functions for cross-functional data topics

**Data Office**

- **Easier for Business**: Data Management as the advocate towards Software development

**Figure 12-3.** The facilitator between business and IT

# Opt-out attitude

Not everybody is either your friend or your enemy. Many are somewhere in between, and a lot of your colleagues may have not decided yet. They will watch for a while before they decide whether to support you or not.

What do they think? In simple terms, something like "If we like it, we follow. Otherwise, we will continue to do our own thing."

Are they forced to support and follow you because you have a mandate?

- Formal Board support is great – but enforced loyalty is not as effective as voluntary loyalty.
- Convinced sponsorship of individual executives is great as well – but it is fragile in nature: Managers come and go.

Please don't rely entirely on support from the top! Even if you have it, prepare for a time without it while making the best possible use of the initial support of those who took you on board.

There are departments in any organization that do not depend on being loved. But the Data Office is not the Legal department or InfoSec! All teams should *want* to work with you.

But how do you develop voluntary loyalty, and how do you convince the undecided that it is desirable to be on your side?

The key to solving this challenge is to make people want to engage. They need to become accountable, and they need to expect disadvantages of not playing ball.

My recommendation:

**Work on overcoming the "us-versus-them" trap.**

Remember what I said in Chapter 1 when talking about "Centralistic Data Governance"? The degree and direction of a person's loyalty are often revealed through tiny words, and one of them is "us."

If you listen carefully to a person, you will find out what that person refers to when saying "us."

You will find that people often refer to their own department or team. This is a red flag of silo thinking. It indicates their intent to optimize their part of the organization instead of finding the overall optimum.

Employees frequently use "us" to describe the workforce, while "them" are the executives. This view also suggests an imminent conflict of interest within your organization.

Good news is that you can influence such a situation.

---

**Note** This is not about changing the language (which would cure the symptoms only)! It is about encouraging everybody to focus on the **entire** organization.

---

You do not need to ask for altruism. You would instead want to make people think "If it is good for the organization, it is also good for me!"

And being a good contributor in the eyes of the CEO (who represents the entire organization, after all) is a totally acceptable motive for them as well.

In reverse, you can propose data-related performance indicators (ideally even those tied to bonuses) in a way that people get rewarded for better data and for data-related successes in *other* areas.

A CDO and a Data Office which, by definition, have the entire organization in mind rather than only parts of it benefit immediately from such a perspective.

As soon as a business colleague (subconsciously!) starts using the word "us" in the meaning of "you and me," you might have found a promising collaboration partner.

Ultimately, among executives, you should seek collaboration with those who frequently use the word "us" to refer to your entire organization – while "them" denotes its competitors.

# Disengagement

*Let's sit down and watch how the CDO fixes all of our data problems…!*

Isn't it a comfortable situation to be in? You can lean back and watch someone else try to complete a mission. If that person is successful, you can congratulate – if not, it is that person who will be blamed, not you.

A certain number of disengaged colleagues is okay. But how do you react if you face too many disengaged peers? Maybe disengagement has developed as part of the organization's culture, as those who engage are tasked with implementation, or as engaged managers are perceived as "junior".

As a rule of thumb, make people accountable, and reward those who are supportive.

The following steps could help:

- Share the fame of success publicly with those that were with you (and who deserve it!). It must pay off to join forces with you. Praise does not cost you anything.

- Concentrate the energy of the Data Office on activities supporting those peers who engage. Unless you have a tremendously big team, this will be enough to keep you busy!

- Use your authority (and governance) to allocate responsibilities in data matters to peers. And use the mechanism of data ownership. People who disengage may end up with a devastating record, visible to everybody.

- Invite known supporters to your governance bodies. Let those bodies decide about responsibilities in dealing with data. Make people want to be called onto those boards, in order not to leave the decisions with others. Make them understand what it takes to get there.

- Propose relevant data-related roles to disengaged peer: "Given your expertise in both Analytics and Marketing, I would like to suggest (read: promote) you as the organization's Data Champion representing the entire Marketing department!"

- Demonstrate that "data" is future-proof and that it has come to stay. Describe the opportunities of early joiners. Make people not want to miss the train.

# Skepticism

*Yet another layer of bureaucracy…!*

People have seen many hypes come and go, promising significant improvement. Most of them did not deliver, but they always added complexity.

A new, dedicated data management entity may provoke the same concerns. You will typically hear statements like "Why can't I go straight to IT anymore? If a data office sits in between, everything just takes longer!" or "Yes, data is important. But what is the added value of yet another function?"

Such a perception is understandable but dangerous. If not addressed adequately (and quickly!), it will cause people to ignore the new Data Office and to continue to work as they had done before. Such behavior has rendered entire Data Offices useless and marginalized them.

You cannot address such a situation through conventional training. Here, people do not primarily need to *know* – they need to *see*.

But how to make employees on various levels and in various roles "see"?

If you cannot create immediately visible value through the execution of typical Data Office activities, do your colleagues a favor or two!

In other words, listen to their concerns – beyond data! Help them wherever you can. Lend them a project manager to set up their project effectively. Have your Glossary Manager review their documents from a language perspective. And so on.

This approach might seem to unnecessarily divert energy from your "true tasks." But, believe me, this is energy well invested!

# Business arrogance

This may be a bold headline. But whoever has had a colleague reject help knows how it feels to see people fail who thought they can do it all on their own.

And, yes, sometimes it is difficult to accept the offer for support – even if you are struggling. It is your own pride, perceived expectations from others, or the ambition to "have managed it all on your own."

Even if such behavior is not in line with the obligation to do the best for the organization, nobody will give it up easily.

You can take a bilateral and a multilateral approach to resolving situations of this kind.

The **bilateral** approach seeks personal dialogue with a colleague who rejects your support. The first step is to find out the underlying motives of the observed behavior.

You can then tailor your strategy on what you learn about your counterpart:

- You may offer to keep your contribution in the background so that the colleague still looks successful. Or you propose a formal partnership that demonstrates both target achievement and collaboration.

- You could agree on a deal, for example, where the first phase of a business initiative will be completed without support from the Data Office, while a second phase will have a data expert be a member of the team from the outset.

- A third option could be to ask for help in reverse to avoid a situation where someone owes you a favor.

The **multilateral** approach aims at establishing a culture that encourages mutual support. This can get to a point where someone looks bad, who does not reach out for help.

This option is usually recommended in case too many individuals are genuinely convinced that they know better or that their traditional methods are superior to a data-driven approach.

Where these people form a small minority, however, you might not need to do anything. Let them fail and concentrate on supporting others.

# Summary: Prerequisites for success

Going through all of these challenges, we have seen a lot of approaches to mitigate or fix them.

But it is also fair to state that a few preconditions for success are out of control of a Chief Data Officer.

If you don't manage to tick off the following three prerequisites, you will probably struggle throughout your endeavor as a CDO.

# Board sponsorship (active!)

You need real support, not sympathy.

"This is the right thing, without any alternative. I will defend and push it" will work.

"Looks nice. Go ahead and try to create value…" will not allow you to be successful.

Don't expect any Board member to be perfectly prepared to support you – this is a rare exception. Even those who created your position may have fulfilled a perceived obligation: "As a modern organization, we should have a CDO." This, however, doesn't guarantee the necessary support.

Next to enabling your team to work, or to creating a team if you start from scratch, I'd recommend that you spend most of your time during the first 30 days on individual dialogue, first with colleagues dealing with data on a day-to-day basis, then with Board members. Ask questions and determine pain points. When talking to executives, always explain, illustrate, and outline the risks of not having full Board support.

You will not need all Board members to wholeheartedly support you. But you require key players by your side. Remember, they don't need to understand the details of data logic and algorithms. But they must be convinced that a CDO and a Data Office are a necessity, and they must believe in you to be the right person in that role.

After 30 days, you might wish to schedule your personal review of the situation: Do you feel comfortable? Can you count on enough Board members if any of the challenges described materializes? How confident are you that the Board will support you, or at least not support any of your opponents?

# An adequate reporting line

In case you as a CDO report into the CFO, put yourself in his or her shoes: Is the CFO going to focus primarily on solving Finance issues or on supporting customer service, sales, and production?

The answer is simple. If a CFO saves the world but does not fix the Finance challenges, the verdict will be "failed!"

That is why, in such a situation, it is not the executive who needs to get blamed but the organization's reporting and rewarding structure.

Functional executives deal with functional problems – cross-functional executives deal with cross-functional problems. It is as simple as that.

As a consequence, a CDO should report into a cross-functional executive or into a functional executive with an explicit cross-functional mandate.

Unfortunately, most executive roles are functional ones – be it Marketing and Sales, be it Operations, HR, or Finance. Some organizations even have just two employees with cross-functional roles: The CEO and the CEO's personal assistant.

And, in fact, reporting into the CEO is not the worst position for a CDO, given the criticality of the topic. Many Supervisory Boards, however, want a Data Office to prove its value and to justify its criticality before they enhance the Board by a CDO. The hurdles are still high for a CDO on the Board.

In such a case, there should be a Chief Transformation Officer, a Chief Innovation Officer, or a similar, explicitly cross-functionally mandated position for the CDO to report into.

If none of those roles exists in an organization, the CEO can explicitly give one of the Board members an additional, cross-functional mandate, sufficient to manage data in addition to existing, functional responsibilities.

Don't settle for less! This is not about you being as high up in the ranks as possible – it is about you being able to do your job.

*"Who cares what scientists think?!
Our point of view is much more
popular on social media!"*

timoelliott.com

**Figure 12-4.** An example of an inadequate reporting line

## Clear expectations

You are the new CDO, so you can be expected to achieve global peace, remove poverty, and prevent earthquakes.

Right?

While this is, of course, tremendously exaggerated, the range of expectations is vast. It reaches from "At least it will not do any harm…" to "We expect those 15 people to turn us from a brick & mortar corporation into a digital enterprise."

So your target would not be to raise or to lower expectations. Instead, you'd strive for an ambitious yet realistic view on what a CDO can and should achieve.

Furthermore, you should tie expectations to assumptions: "If XYZ happens, then we can achieve ABC." This is important for two reasons: It makes you less dependent on external factors which you cannot influence. And it allows you to state your financial, organizational, and strategic needs up front.

Very soon, you may wish to put together your *conditio sine qua non*[1], that is, all the conditions without which you are not willing to make any commitment.

These conditions may range from your budget and staffing situation to a minimum degree of necessary organizational and strategic freedom.

---

[1]While the correct Latin plural would be "conditiones sine quibus non", the expression is usually quoted in its singular form.

You will certainly be able to gauge how much of this should be asked before you accept the job offer – but it may not be wise to wait too long for the rest.

While you would spend the first 30 days on securing executive support, it could be the right timing to present your needs during the second month, in conjunction with your high-level objectives and plans.

But don't spend too much time detailing out those plans. Instead, you should already start focusing on the first low-hanging fruits as you might strive for a positive performance record of your first 100 days as a CDO.

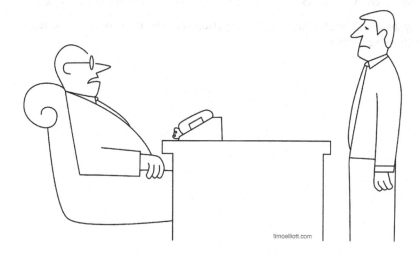

*"The only Big Data letters I care about are the four Ms —*
*Make Me More Money!"*

**Figure 12-5.** Data – what the Board really cares about

## Clear roles in data matters

You may want to have early clarity about your rough set of responsibilities. The latter helps avoid being blamed for issues in other areas. Your CFO should understand that your team is not responsible for a misconfigured data lake or for delayed access to critical data sources. It may look like an excuse if you clarify such things only after one of your teams gets blamed for an overall delay of a complex business project caused by somebody else.

A new function such as a Data Office inevitably changes the delineation between functional responsibilities.

Next to the organizational need to avoid gaps or overlaps, you should use your first days to prevent internal competition, for example, with the IT teams, with a possible digital team, with business transformation teams, and so on.

You should also use this time to outline your ideas about data-related decision and collaboration bodies, about who should be invited and what their tasks and authority should be.

While your detailed organizational setup will develop over time, you might wish to use your first 60 days to address these fundamental sources of conflict. This will allow you to concentrate on the relevant content topics later.

And again, this is not about increasing your area of responsibility – it is about managing expectations.

# How (Not) to Behave As a CDO

*See no evil, hear no evil, speak no evil, tweet no evil...*

**Figure 13-1.** The four secrets of a leader?

© Martin Treder 2020
M. Treder, *The Chief Data Officer Management Handbook*,
https://doi.org/10.1007/978-1-4842-6115-6_13

# Don't rely on formal authority

Of course, you should insist on starting with sufficient formal authority – but don't rely on it: Yes, everybody should know that the Board supports you. This, however, is a "necessary condition," not a "sufficient condition."

That is why, in many cases, you will have to work with informal power.

To say it with the words of Monty Halls,[1] you are facing "the moment in life where you need to lead." You cannot administer, delegate, or manage.

# Start small, and pick your battles

You don't need to double your organization's revenue within your first 100 days. Instead, ask yourself:

- What are the most significant pain points?
- What are the areas where data is dealt with (most) inefficiently?
- What are the low-hanging fruits?
- Which areas has nobody else been willing to own so far?

The important criterion of your first actions is not "efficiency" but "effectiveness." In other words, the absolute result counts more than the ratio between result and effort.

A piece of good advice comes from Abhijit Akerkar, Head of Applied Sciences at Lloyds Banking Group. He says: "Don't start where you have most data – start where the impact is best."

But don't understand this as a suggestion to act tactically by all means. To avoid dirty workarounds, you should always address even your first topics in a way that the solutions can be applied beyond the concrete case later.

# Be humble

Don't try to appear as the one who knows it all. And don't pretend as if your mere role gave you the right to tell people what to do.

As warned by Mark Coleman from Gartner during their 2017 Data & Analytics conference in London, "Trying to assert control leads to resistance, not cooperation."

---

[1]Monty Halls is the founder of the education management firm Leaderbox, creator of wildlife and adventure documentaries, and a famous TV speaker.

But what is the alternative?

- Instead of saying "From a data perspective, you need to do it like this," you could ask "Would you like to know the implications of your approach on data?"

- Instead of saying "Data will do better what you are doing today," you could say "Data will support you in doing your work even better."

- Instead of saying "You need to get your duplicate data under control," you could say "I am sure you have your duplicate data under control. Would you share with me how you do it?"

Note that the third question is not meant to be a trick question! If duplicate data is NOT under control, your counterpart will have to admit it, and you can offer your help. If it is, in fact, well managed, you can even learn something. And you could appreciate the approach and ask whether the two of you could jointly introduce that concept to other areas of the organization.

# Present yourself as a facilitator

Not all data activities should be centralized. Instead, all functions dealing with data should understand data. Your Data Office would provide support in the following areas:

- Creating the foundation: The basis business teams can work on.

- Providing a governance framework that can adjust to different business needs.

- Providing training and education.

The target could be to make your colleagues' lives easier so that they can concentrate on their core jobs. (Remember, most of them only do some of the groundwork because it is indispensable and nobody else does it.)

Example Analytics: If business functions have their own Analytics teams, you may not want to take it away from them. Instead, your team takes care of providing the basis, that is, clearly defined "single source of truth," unambiguous verbiage, and a standardized data model across the enterprise. This allows these teams to dive into the data to gain insight from their functional perspective. They don't need to take care of all the groundwork anymore. Remember that you need to find a balance between organization-wide truth (predigested) and the freedom of functional Analytics.

Remember that you may *offer* to take care of a department's Analytics task. Some departments might be happy to accept your offer (which might allow you to create an even stronger Center of Excellence for Analytics). People appreciate having a choice. Give them this choice wherever both options are acceptable from a data management perspective.

Example Reference Data Management: Business functions should maintain the data they own. A Data Management team may not even need to have an approval role here – as long as the rules are described, the approval processes defined, and so on.

# Avoid suboptimal language

The language you use with your team is different from that of your internal customers. The Head of Marketing may not know what an entity-relationship diagram is. Content-wise, however, the desired relationship between two entities will be immediately apparent to him or her if explained in nontechnical words.

Moreover, you should take the perspective of the person you are talking to. It is different from your data perspective. Compliance with the Corporate Data Model has no relevance for a businessperson – they would not see any benefit of adhering to the CDM.

Finally, it is not only about avoiding tech speak.

Which words do business folks *not* want to hear, despite the content being relevant to them? This list is different from organization to organization, but these words almost always make it onto the list:

- Data
- Governance
- Rules; compliance
- Wait; long term
- Data model (including the related verbiage such as object or cardinality)

It would be helpful to determine the list that matches your organization. Please think of alternatives to all of these words and formulations. Explain to colleagues and ask them how they would call it. Test any alternative terms with colleagues before you introduce them.

# Go out and talk to people

You cannot effectively turn a traditional organization into a data-driven organization if you don't walk (or travel) around and talk to people.

Or, as Ken Allen, who turned around DHL Express as its CEO, put it: "One thing is for sure: You can't build an outstanding business from behind your desk" (Allen, 2019).

Data management is about trust more than it is about anything else.

How do you gain trust? Talk to people, face to face. Visit them at their workplace – whether it is an executive office or an assembly line.

And don't start with sharing your wisdom. Listen and ask questions.

It is imperative to establish an interpersonal relationship with people whom you expect to read and follow your messages later. Ask yourself: How intensely are *you* reading messages from people you don't know?

# Stakeholders

*"Err... Do you have any alternative facts?"*

timoelliott.com

**Figure 14-1.** Not everybody likes your messages

© Martin Treder 2020
M. Treder, *The Chief Data Officer Management Handbook*,
https://doi.org/10.1007/978-1-4842-6115-6_14

# Manage stakeholders at all levels

As in every other role, you need to know your standing in the organization to be able to plan your approach. To do so, you will have to

- Know your supporters and your opponents

- Understand both their knowledge and their motives

- Get a good feeling of their degree of support, that is, their willingness and their power to help you achieve your targets

Stakeholder Management is not limited to the top levels. Virtually everybody in the organization is a stakeholder. (Have you seen an area that does not deal with data?) Data Management cannot (and should not try to) hide in a cozy corner, just dealing with their neighbors…

This chapter is meant to share ideas about effective shareholder management. Most thoughts are not specific to data management, as all of this is about dealing with humans, across all functions and all hierarchy levels.

## Document your insights

You may have been with your company for many years, and you may be convinced that you already "know" most of your stakeholders reasonably well.

But do you know what drives all of them? Are you sure about their reactions to concrete proposals?

For truly effective stakeholder management, you will have to put in some extra effort. And it is of the type that we often try to avoid: Tedious documentation. But I promise that it is worth the effort! Here's a quick summary of my favorite, three-step approach:

1) Systematically maintain a (potentially huge) stakeholder list. Cover all functions and all geographies.

2) Separate your stakeholders into those who are positive, neutral, and negative toward your ideas and vision.

3) Document whether someone is **actively** supportive or inhibitive. Those people require special attention.

## Classify your stakeholders

Your most loyal supporters can usually be found among those unhappy with the status quo, particularly with the problems you intend to address as a CDO.

So, watch out for those as you go around and sell your story.

Hold on – there is a second criterion! You don't want the whiners nor those who feel good complaining about everything. You are after those who are ready and willing to act. Alas you have to expect them to be a minority (see Figure 14-2).

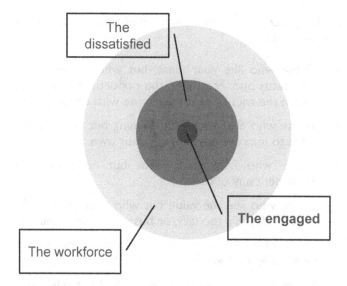

**Figure 14-2.** Finding your supporters

These are the people you want to involve in all discussions, giving them the (correct) feeling that it's *you* whom they have been waiting for, for such a long time.

How do you determine those supporters?

a) Determine the "dissatisfied" and the "visionaries."

- If there is a problem, there must be folks who suffer today – find them! Ask yourself up the hierarchies, from the people who told you, up to the highest manager who consider the current situation a problem.

- Even where there is no problem, there may be people who understand how data could help improve the current situation – find them! You may find more experts than managers – that is okay!

b) Determine the "engaged."

Not all dissatisfied and visionary people are ready to support you actively.

You may present your vision, and everybody agrees. Does this mean you have broad support now? Not necessarily, unfortunately.

You will need to do some sort of gold panning to finally get to the nuggets – to those few who will support you actively.

In this process, you would identify and put aside the following groups:

- Those who like your ideas, but who can live with the status quo. For them, the opportunities are not worth the friction that may come with engagement.

- Those who are always complaining but without any drive to improve anything on their own.

- Those who have great ideas but no energy to implement any of them.

- Those who see the value but who want **you** to do it all as they are too lazy, or busy with other things.

- The cynics who mock about the resistors as they want to feel superior.

- Those who pretend they understand but who have no clue.

- Those who wish you success but don't require any such change for them to be (considered) successful.

What remains are the gold nuggets – a usually small fraction of the dissatisfied or visionaries who are ready to act.

I am talking about those who openly support your ideas against resistance, who commit time or resources to your initiatives, and who share the risk of failure with you.

They are the people you are after – across all hierarchy levels.

## Determine your executive allies

A powerful ally is better than an engaged but toothless tiger, of course.

That is why you should have a particular focus on those whom you want (and need) to be your allies or sponsors.

It will usually be a tiny number of influential executives. These relationships deserve a significant share of your time and attention.

If you don't enjoy the luxury of being a Board member yourself, you need an Executive Sponsor. And it must be an active sponsor! As Brent Dykes[1] once put it on his blog:

*The difference between success and failure often comes down to a sponsor's level of commitment and involvement. Without commitment, an involved executive sponsor may attend every meeting but only provide lip service—never backing up the initiative with the resources, budget or political influence it needs to be successful. Without involvement, a committed executive sponsor becomes an absentee cheerleader who is too busy to contribute to the initiative's success.* (Dykes, 2016)

## Know the motives of your allies

There can be various reasons for people to be on your side. That is okay. But you should always know *why* someone wants to support you.

- Those that want to do it "for the greater good" are usually the most reliable supporters – but rarely the most influential ones.

- Those that think your enemies are also their enemies are dangerous supporters. Their view may change at any time, for example, when people move to new roles.

- There may be those who expect you to solve their self-made, possibly data-independent problems. Here, you will need to manage expectations while still confirming that you will make their lives easier.

All in all, please consider that your supporters are not identical to "good" people, just as your opponents are not necessarily "bad." Some of your colleagues may oppose you with good intent, while others may support you for selfish reasons.

## Concrete recommendations

(i) **Create allies, not subordinates.**

Nobody should follow you because they are forced to, let alone from fear.

On the other side, most people won't support you just because you are a nice person.

---

[1] Brent Dykes is the Director of Data Strategy at Domo, a cloud software company based in American Fork, Utah, United States. It specializes in business intelligence tools and data visualization.

It is, therefore, essential to focus on value creation, to search for a **joint basis**, and to shape win-win situations.

Pilot projects help a lot. They allow you to show tangible results quickly, which increases buy-in.

### (ii) **Invest time and energy.**

I already mentioned that it is crucial to understand why single players act the way they do. Don't accept easy answers! You may, at times, require a complete root cause analysis to get to the bottom of someone's motives.

### (iii) **Work with the "engaged."**

You don't just "have" supporters – you need to involve them actively. Here are a few ideas:

- Work with supporters on making a case. It needs to be a joint case, not "your own" case.

- Create virtual teams of those who are ready to act in your sense.

- Give them a voice. Don't expect them to slavishly follow your own ideas.

- Agree on a shared vision. Develop an actionable plan together with all of them.

- Give people a (formal or informal) role, so that they become a regular part of your endeavor.

### (iv) **Maintain the dialogue with Board members.**

Board members should always be aware of the need to act. Suggest a recurring meeting schedule with individual Board members, or agree to touch base every now and then, so that you don't end up being a supplicant.

### (v) **Make people provide resources.**

If you need funding, join forces with a funded initiative: Try to create a win-win situation where they benefit from having you on board. Such a situation often motivates project leads to share their budget and resources.

Furthermore, it is worthwhile determining and exploiting voluntary IT resources: Those who like your ideas and who have some bandwidth to support you, outside of any formal resource allocation.

(vi) **Mediate, mediate, mediate.**

Don't underestimate the power of balancing between varying concerns. Talk to individuals, talk to pairs of individuals. Listen! Add different dimensions to negotiations to avoid a "slice-the-pie" situation where people get tied to their own "red lines".

A successful mediator wins on both sides.

(vii) **Actively create trust.**

Trust can hardly be gained through tricks – it must be truly deserved. And you should demonstrate through your behavior that you are a trustworthy person:

- Be predictable.

- Promote organization targets, not your own.

- And keep your promises.

(viii) **Never shoot back.**

Your opponents will try to harm you. Don't respond in kind. Outsiders will not be able to tell who's right and who's wrong.

# Tailor your stories to your stakeholders

Somewhere in the Orient, about 500 years ago, the ingenious Omar has invented a new teapot, using a new type of clay, shaped in two layers with a layer of air in between. Due to prefabricated molds, production would be efficient and cheap.

He decides to sell it to the Sultan. He gets an audience, and he tells his story: "Look, this teapot has a huge capacity, it keeps the tea hot for a long time, the handle is very comfortable, and production is cheaper than that of conventional teapots!" – "So, what!", says the Sultan, "I get tea whenever I ask for it, there is always enough tea, my slaves keep it hot, and I never need to touch the handle. And compared to my wealth, the small savings won't make too much of a difference. I am not interested! Next, please!"

Back home, Omar thinks it through again. So many tangible advantages! The Sultan just didn't get it – what a fool! And if he remembers me, I will never be able to sell him anything again in future.

Omar decides to relocate to another Sultanate and to try again there. This time, however, he chooses a different strategy. He understood that he should not blame the Sultan but his own, inadequate sales pitch.

As a consequence, he determines relevant stakeholders at court, including the chef, the supervisor of the Sultan's personal slaves, the cupbearer, and the treasurer. Then he makes appointments with each of them.

He explains to the chef that his insulated teapot doesn't require constant heating which keeps the aroma stable for a longer time than traditional teapots. The treasurer learns how much money can be saved per teapot (and that many teapots save a lot of money). The supervisor is given a chance to test how user-friendly the teapot is and that the handle really doesn't get hot. This would save his slaves a lot of trouble. Later, the Sultan's cupbearer is invited by Omar to a cup of tea. Omar shows him how easy to use these new teapots are that he produces and sells in his pottery.

A few days later, during a meeting the Sultan has with his closest advisors, a teapot smashes on the ground as a slave burns his fingers and drops it accidentally.

The angry Sultan calls for his slaves' supervisor and asks him to ensure that such things don't happen in his presence. The supervisor apologizes, and he mentioned that a new shop called Omar's Pottery has a teapot that doesn't get very hot on the surface. The cupbearer adds that Omar's teapots hardly ever spill any tea.

The chef who is supervising the food supply during the advisors' meeting confirms that the teapots from Omar's Pottery keep the aroma, avoiding any bitterness of the tea.

The Sultan says "Omar's teapots must be fantastic! They probably cost a fortune, but they may be worth the investment!"

The treasurer jumps in, stating "If I am not mistaken, these teapots are even cheaper than the ones we have been purchasing so far."

"Why on earth," says the Sultan, "haven't we bought any of these teapots yet?" Soon after, Omar's Pottery came to be known as "Purveyor to His Majesty the Sultan."

What can CDOs learn from this tale?

a. **It is not sufficient to have all the right arguments on your side.**

b. **Different aspects are relevant to different stakeholders.**

"...*and the customers, shareholders, and employees lived happily ever after!*"

**Figure 14-3.** Tell a story that resonates

## Ask the right questions

Why is it so important to ask questions and listen to stakeholders before coming with stories or proposals?

You may have guessed it already: Before creating an adequate story for a stakeholder, you need to understand this stakeholder's position.

Please expect everybody – not only executives – to ask the question "What's in it for me?" You need to be prepared to respond convincingly. And different stakeholders require different answers.

But how can you develop a perfect story for each stakeholder? In the first step, by concentrating on any gap that the stakeholder complains about or at least agrees to.

This allows you to pick the one or two of these weaknesses that cause severe pain to that stakeholder or for which you think you can offer the most attractive solution.

You may wish to start with a list of current weaknesses in your organization which a data approach can help overcome. As a second step, you'd check by key stakeholder how they see each of these gaps:

1) Do they share your judgment?

2) Would they do so if you explained it to them?

3) Do they think they can solve it on their own or through the current organization, respectively?

4) Or would they deny the existence of the weakness (maybe because they are part of the problem)?

This (living!) document – a simple matrix would do – should become the basis of your active stakeholder management.

Your story would then include a root cause analysis: What causes this weakness? As we know from Six Sigma, you usually need to ask "why" five times on average before you arrive at the ultimate root cause(s).

Whenever possible, you would spice your story with real-world examples. When talking to executives, it may resonate well if you use cases you learned from members of their teams.

Here are a few questions you can ask any Board member:

(i) **Awareness**

- Do you know how good our organization's Data Quality is? Be it for operational purposes, be it for Analytics?

- Do you know what data we have?

- Do you know what data we could have?

- Do you know the value?

- Do you know the opportunities?

- Who ensures your data is managed consistently across the organization?

- How quickly do you find out about a communication disaster for your organization? Early, by analyzing social media, or late, through the press?

- Can you relate a negative tweet on Twitter about your organization to a particular incident involving the tweet's author?

(ii) **Status questions**

- Do we have the right people?
- Do we approach data the best way?
- (How do YOU deal with data?)
- Are we ready for data?
- Do you feel confident today?

(iii) **First ideas for the way forward**

- How about a data strategy, to support the corporate strategy?
- How about cross-functional collaboration?
- How about creating awareness among ALL employees?
- How about regularly measuring the quality of our data?

# Pick the right weaknesses

The list of weaknesses itself is strongly dependent on the situation in your organization. That is why it is essential to focus on weaknesses from a *business* perspective.

An issue like, say, "Lack of data ownership" is therefore NOT a weakness you'd put onto this list. It is rather one of the root causes you'd derive from weakness as part of your story.

So, what are typical business weaknesses you can address through your data approach?

In general, you can distinguish between executive concerns and daily business concerns.

Typical daily business concerns are

- I lack information. And there is no systematic way to find what I need.
- I usually find out if something goes wrong, but I don't know where and when it happens.
- If I want to change something, I have to submit a request months before the actual change.
- I often find that I work with outdated information (and usually I don't even know whether it is still accurate).

- Our IT colleagues don't understand our business.

- Our business colleagues don't understand basic IT principles.

Executives' concerns could be

- We don't discover undesirable development early enough.

- I would really like to base my decisions more on facts. But in many cases I simply don't have all relevant information, and sometimes information comes in a form I cannot use without investing too much time.

- We do not have the necessary flexibility to react to changes in the market.

- Business changes (e.g., introducing, changing, or removing a product) take too long, involving various departments and a lot of manual work.

- We seem to have interoperability issues with acquired organizations (or after a merger).

As you can see, the word "data" doesn't appear anywhere on this list – because business people don't think in terms of "data."

Don't blame them. Honestly, it is the CDO's responsibility (and opportunity!) to link business challenges to data issues and subsequently to data solutions.

When listening to your stakeholders, you will probably determine a short list of weaknesses that appear again and again. However, they usually materialize differently in different business functions.

As positive as it may be to get the opportunity to present your story in front of the entire Board, this should only be the last step after you have guided most of the Board members through "their" individual journey. You cannot address everybody's particular pain point while keeping your presentation crisp and clear.

And, remember, you are not expected to present different realities. It is the focus that differs from stakeholder to stakeholder. An overall story on Board level must never contradict any of the individual stories. It should instead signal that each of those stories fits in very well with your data story for the entire organization.

# Keep data on the agenda

Don't present once and run away. Executives tend to have a short memory even if they seem fascinated. What feels like the top topic for you personally might be agenda item 7 out of 16 for the Board members.

Here are a few recommendations to keep the ball rolling:

(i) **Ensure you get invited again.**

It does not look too good if you ask again and again for permission to present to the Board. Instead, you may wish to make **them** ask you to come back.

The best opportunity is the moment you present. Tell them a lot is going to happen in your area, and ask them whether they'd like to be kept up to date on your progress.

Offer a concrete recurring update (but be modest – you cannot be expected to become part of the standing Board meeting agenda!).

(ii) **Create a regular data topic.**

Even if you don't attend all Board meetings, data may make it onto the list of recurring agenda topics.

A promising approach is the definition of a Data Scorecard. Metrics are the shortest way of telling people where they stand, and Board members love shortness.

Maybe a Board already has a list of Key Performance Indicators (KPIs) shared regularly. Typical examples are customer satisfaction or revenue by channel. Try to have the "data" subject added. Maybe you even find a catchy name like "Data Dashboard" or "Data Compass."

Ideally, you'd have a hierarchy of Data KPIs (which I call Data Quality Indicators, in short DQIs – see Chapter 10's section about Management of Business Metrics) that you work with anyway. Share the top-level DQIs with the Board, and only break down where there is a good reason.

But don't focus on issues alone – share relevant improvements, and describe the impact on your organization's success (revenue, market share, reputation, etc.).

All of this can be put into a five-minute update. The key message should be "We have it all under control" (which you hopefully do). Based on this message, make them curious. Next to selfishness ("What's in it for me?"), it is their curiosity that keeps executives engaged.

# Shape your data network

Here is an example of what the different elements of an organization's data network could look like.

## Functional Data Champions

A lot has been written about Data Champions, with the result that there are as many opinions about their roles as organizations are working with Data Champions.

My preferred setup is that of Data Champions by unit (e.g., geography or entity) combined with Data Business Owners who are the business custodians of a specific subset of the data model. Each business function and possibly each geographic entity would select one Data Champion to represent the interests and needs of that group in the area of data. This is usually an additional role for a seasoned, data-savvy functional expert.

The role of a Data Champion could entail

- Being the single point of contact toward the network, that is, for cross-functional data topics from the Headquarter and all entities.

- Being the internal "single point of contact" for any data-related question, inquiry, or proposal from any team member. Share and explain the organization's Data Principles as well as any messages from the Data Office. Ensure everything is understood within their area of responsibility.

- Local data network: Knowing the local Data Stewards, so that you can bring them in contact with each other. Helping them comprehend how their work impacts the work of Data Stewards in other functions.

- Ensuring local expertise is brought into organization-wide discussions, for example, knowledge about local address handling habits, best sources of external data, or local legislation impacting data handling.

- Taking care that functionally or geographically focused projects explicitly consider data aspects and support them from a data perspective. Collaborating with the central Data Management team for a consistent approach.

- Understanding local, divisional or functional controls to ensure Data Quality on a day-to-day basis.

- Observing local activities and looking out for data-related issues. Running local initiatives and root cause analyses based on data findings, using Six Sigma methodologies.

- Consolidating local needs with regard to data (requirement toward tools, e.g., analytics visualization, data sources, country-specific data structures, e.g., for required trading licenses; validation of local data sources, e.g., in case of statutory changes).

- Determining best practices (as well as approaches which did *not* work well) through the network, in preparation of own or supported local or functional projects.

- Facilitating the discussion where team members from different functions have different views on how to handle data locally.

# Business Data Owners

Data Ownership does not come naturally. Most members of business teams are not used to being responsible for data.

But as data is cross-functional by nature, you need someone to coordinate between all business functions in case of suggested changes or challenges for each possible data element – the Business Data Owner.

### (i) **A Data Owner**

Each Business Data Owner represents a subset of the Data Model, for example, one Data Domain (e.g. "Customer" or "Product"). He or she has to align with all stakeholders of that Data Domain, beyond their own business area.

This is usually not a dedicated position. For this role, an organization requires open-minded people who don't wear blinkers. They need to be able to foster data collaboration between business functions.

### (ii) Data ownership

The required definition of data domains should be coordinated by your Data Architects. They may follow existent de facto standards or best practices, but they should always take the characteristics of your organization into consideration.

Sometimes Business Data Ownership strictly along the borders of Data Domains is not optimal, for example, where different departments are responsible for different attributes of one entity within a domain. In this case, you may wish to deviate from your organization's Domain Model. This is easier if your Domain Model is already broken down, for example, into business objects, subdomains, entities, and/or attributes.

### (iii) Determining Data Owners

But how do you practically agree on Business Data Ownership? I usually recommend a three-phase approach:

Phase 1: A Data Office team determines the ideas of all stakeholders, ideally in a personal dialogue. They would use the Domain Model as a template to guide the potential Business Data Owners through all domains, and they record proposals. Wherever there is a single volunteer, for example, for a domain or a business object, it is an easy tick in the box.

Phase 2: The team proposes Business Domain Owners for all uncovered areas based on what they have learned about the business during Phase 1. The overall result is cast into a first comprehensive Business Data Ownership matrix, i.e. without any blank spot.

Phase 3: The Business Data Ownership matrix is sent to all stakeholders with a request for review. It is critical to provide strict review rules to avoid an uncontrollable mix of personal opinions:

- The reviewers are not asked for their preferences or points of view.

- Instead, they are expected to state where they disagree and, most importantly, *why* they disagree.

- Only challenges with a valid business reason can be considered.

- Whatever is not challenged is considered as accepted. Lack of (timely) feedback is regarded as full agreement.

All noncontradictory feedback gets incorporated, and disputed ownership points are marked as such.

The result will be channeled through the (cross-functional) data decision bodies, asking for decisions for the remaining points of dispute.

As, by design, the Data Office has no stake in any individual allocation of ownership, you can accept any of those business decisions without jeopardizing your data mission.

# Data Creators: Data Stewardship Network

These people have the authority to maintain data on demand – be it the insertion of a new country code or the modification of a customer record. They are usually spread across departments.

As the work of Data Creators has an impact on other departments, they should form a community for an ongoing dialogue. People are not "appointed" as a Data Creator – they are Data Creators by virtue of their regular business role.

# Data Consumers: Analytics Network

Business functions often have their own, self-contained Analytics teams. While it is good to have people understanding both their business area and data, this setup has led to a myriad of challenges:

- A multitude of different, incompatible solutions, each coming with its own technology and logic.

- Different ways of interpreting the same data.

- Lack of transparency with regard to data sources, applied logic, and others.

- No single source of truth.

- Unmanaged data lineage.

- Ungoverned usage of data by too many different users.

- Hidden gems: There is a lot of valuable data and logic in the organization that people are not aware of.

In response, the Data Office and IT may jointly set up an Analytics Network for all business functions. Here are a few typical topics for this network to deal with:

- Target architecture (led by IT)

- Technology and operations (led by IT)

- Shape a joint governance structure for Analytics (led by the Data Office)

- Agree on collective approaches to measure and improve Data Quality (led by the Data Office)

- Collaboration: Exchange of experience between teams (led by both IT and the Data Office)

## Double loyalty

How to make people want to be part of a Data Stewardship Network or a Data Analytics Network? After all, they sit in different departments and report to different bosses.

You certainly don't want to replace functional loyalty with cross-functional data loyalty. Good news is these two kinds of loyalty are not mutually exclusive.

People can keep their functional loyalty through formal reporting lines, and they can express their commitment to the world of data by becoming part of a data community (see Figure 14-4). As a result, they can be the ambassador of each of the two worlds when interacting with the other one.

**Figure 14-4.** Data experts should feel loyal to both worlds

# Orchestrate your data network

It is an excellent opportunity for a Data Office to orchestrate the data network actively.

Online collaboration platforms such as Yammer are great for this purpose. But a collaboration platform needs to be actively operated. People will use it if they find it helpful, notably if their questions get answered – be it by the community, be it by the Data Office.

This helps reach the "critical mass" which you can reach more quickly in larger organizations. Nobody checks an online collaboration platform that has less than one relevant user entry per day.

A key concept for a collaboration platform is to give access to everybody. While steering committees usually suffer from too many members, a collaboration platform wins with each additional member.

The same applies to information sharing. If people know that there is one single platform where all information is shared, everybody will look there first.

That is why such a platform, once accepted, can also be a perfect channel to disseminate messages and information around data.

Finally, please think of joint initiatives, competitions, and rewards to make people feel part of the data community.

You can formalize membership to make it look more valuable. This would allow for "Members only" conferences to make people feel valued as network members. Little gadgets like "Data Community"–labeled coffee mugs contribute further to such a feeling of belonging.

# Plan to consider different audiences

You will not reach everybody with the same message.

Not only will you need to distinguish between a Board member and an admin clerk – but you will also choose different wording when talking to business folks vs. data-savvy people.

It may, therefore, make sense to also formally group the audience, that is, to set up formal groups of people you can address similarly.

First of all, Data Creators and Data Consumers usually do not have too much to share with each other on a daily basis. If you want to discuss Reference Data maintenance, data consumers will get bored. Maintainers of Reference Data don't care too much about data visualization. This is why two separate networks (or two separate areas within a shared chat forum) usually make sense.

Secondly, it may be necessary to divide Data Consumers further into operational consumers and analytical consumers.

Thirdly, you will usually observe a difference between a "data" view and a "business" view.

- The former concentrates on data-related activities, independently of business case aspects and prioritization.

- The latter looks at data topics from an overall business perspective: Requirements, urgency, opportunities, costs, resources, and benefits.

Experience shows that these two perspectives involve different people and roles so that it makes sense to organize them separately. See Figure 14-5 for an overview.

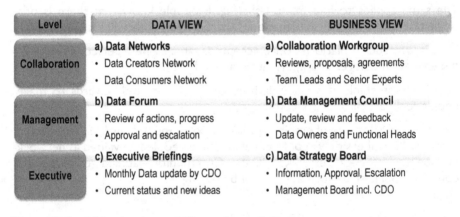

| Level | DATA VIEW | BUSINESS VIEW |
|---|---|---|
| Collaboration | **a) Data Networks**<br>• Data Creators Network<br>• Data Consumers Network | **a) Collaboration Workgroup**<br>• Reviews, proposals, agreements<br>• Team Leads and Senior Experts |
| Management | **b) Data Forum**<br>• Review of actions, progress<br>• Approval and escalation | **b) Data Management Council**<br>• Update, review and feedback<br>• Data Owners and Functional Heads |
| Executive | **c) Executive Briefings**<br>• Monthly Data update by CDO<br>• Current status and new ideas | **c) Data Strategy Board**<br>• Information, Approval, Escalation<br>• Management Board incl. CDO |

**Figure 14-5.** Addressing the two different data audiences

# Frequently stated concerns

The more challenging brother of "Frequently Asked Questions" is "Frequently Stated Concerns." We have touched most of them throughout this chapter, and here is a summary.

## Which problem are you trying to solve?

Executives have their individual priorities and problems. That is why you need a tailored story for each of them.

Let's look at three examples.

## EXAMPLE 1

### The CFO story

A CFO knows how relevant their own role is. So let's ask the CFO: "Who is doing the CFO equivalent activity for data?"

Most organizations have more data than tangible assets. The financial value of data can be demonstrated as well, using examples and reports.

And a CFO will understand if you explain why "Data is an asset."

If data is there, and if it has a high value, the CFO should be the last to deny that somebody high up the ranks should take care of it.

## EXAMPLE 2

### The Production story

The Head of Production may not see the need for active data management as all the KPIs are already there.

An interesting question would be "Do you know the accuracy of your KPIs?"

Maybe the Head of Production sees data similar to electricity: It is simply there, and you just need to plug in a power-consuming device to make it work.

Your questions could be "How quickly would you find out if the data you use is wrong?" and "What would be the consequences of wrong data steering your production process?"

It is crucial to use examples – from both the operative world (e.g., data from a defect sensor) and the commercial world (e.g., incorrect product demand data).

## EXAMPLE 3

### The Legal story

Why should a lawyer in your organization be interested in the management of data?

Interesting questions could be "Do you have an overview of the adherence to our organization's privacy policy? Can we be sure to pass an external audit? Do you have the commitment that no personal data will be kept longer than necessary? Do you know all the people responsible for personal data in the organization?"

Furthermore, you could ask whether the organization, if sued, can quickly pull together all information to prove its innocence (or to find out early that it has a problem indeed)?

There are obviously many more problems that can be solved through data management. It is helpful to maintain a standing list of topics, including all owners (or the people impacted most). You will usually not need to describe how you intend to solve those issues, as most people are more interested in the end result than in the method to get there.

# What's in it for me?

For a Data Office to be successful, anything they do must come with a benefit. Not for themselves – this is too shortsighted. Not for the organization – that is too remote. No, you need to convince individual colleagues that your work improves their situation.

This doesn't mean you need to adapt your targets. Instead, it helps if you shift the focus of expected benefits: Away from valid but abstract data-centric objectives like "high Data Quality" or "single source of truth" toward benefits business teams can relate to.

In general, you'd find out what's on top of people's wish lists – let's call it "A." Then you think about how data can help fulfill those wishes – we call it "B." The value proposition would then simply read "A through B."

Examples are as diverse as "reducing labor costs through automation of processes," "increasing first-call resolution rate of contact center agents through the provision of all customer- and case-related information in real time," and "reduction of overproduction through a precise forecast of demand by day-of-week."

A second component of "What's in it for me?" is the often-underestimated individual aspect: "What's in it for me personally?"

Note that people usually don't state it this frankly – you will have to read between the lines to find out.

In some (ideal) cases, departmental successes become personal successes. Often, however, people long for personal benefits such as less effort in their daily work, better visibility of their own achievements, or protection against unjustified complaints from colleagues.

Finally, you may sometimes have to ask people to follow your direction without being able to derive a (departmental or personal) benefit. The honest answer to "What's in it for me?" would be "Nothing. It's just better from a shareholder perspective."

This is obviously difficult to sell, as most people don't aim at doing good for its own sake, as Aristotle used to suggest.

In such situations, you might have to offer something in reverse. Thank goodness, a Data Office has so much value-add in its portfolio that you should be able to provide something for everybody – be it regular data services or something that the data team has the skills to do, be it to satisfy departmental or personal needs.

# I have no bandwidth for data stuff

I recall a freshly appointed CFO saying to the CDO: "You have a lot of interesting ideas. I promise I will have a closer look once I have time. But first, I need to solve several urgent problems."

In this case, "time" meant the availability of resources but also the necessary attention, which currently was with a lot of short-term topics with immense management visibility.

As, in that organization, the CDO was reporting into the CFO, the Data Office teams could maneuver freely within their supportive role, but strategic development toward a data-driven organization was not possible for the time being.

You can derive two recommendations from this story:

(i) **Data needs to deliver quickly.**

If you offer added value through data, don't concentrate on initiatives with the most significant scale or business case.

Instead, start with initiatives where you can deliver quickly and where not too many resources outside the Data Office need to be busy. This increases the probability of being given a chance to prove your value.

(ii) **Data is not a "work on top" – it helps today.**

Using data to improve business is not a task on top of all the other obligations of a team, thus requiring additional resources. Instead, it makes it easier to solve even tactical problems.

You may wish to concentrate on conveying this message early as it is not naturally clear to executives.

You may recall the famous cartoon where a Stone Age man who has apparently just invented the (round) wheel tries to offer it to two colleagues who are desperately trying to pull a cart with square wheels. This activity requires such a significant physical effort that they don't have the time to listen to the first man's proposal that would make their lives significantly easier right away.

There are many other analogies you could use to explain the opportunities that come with data. However, you might wish to share easily understandable examples from the real world as well.

# We can look at your strategic ideas tomorrow

Have you ever asked an accountant for an appointment while they were busy closing the annual accounts?

No matter how great your strategic ideas may be – in some cases, people have more urgent things to do.

The problem starts where this becomes a steady state. And most people have something to do at all times – from the CEO down to the admin clerk.

Where this is used as an excuse, you might wish to use your data governance to set and communicate the right priorities. But this alone might not help. People won't drop their pen to work with your team right away.

It is usually a good move to allow people to buy time. After all, they don't resist your idea in general – they just don't want to deal with it right now.

So, why don't you make a deal? "You can finish these activities until the end of next week, and then you will ask two members from your team to work with my team on XYZ." Please document this deal even if you do it through a mere email.

Such an agreement will allow your counterpart to get organized and to prepare. And everybody feels the more comfortable, the further the impact of an own commitment lies in the future.

To prevent this approach from delaying your initiatives, you should bring it up as early as you become aware of the need to work with another team.

# IT has always covered this

Here we can distinguish between three different cases:

(i) **Data teams outside the Data Office**

Specific data aspects may indeed have been covered by IT in the past. And if those people or teams have done well, maybe they should continue to do the same job as part of the Data Office. This is a sensitive area where you should listen and act very carefully.

In many cases, previously responsible teams will be happy to hand over ownership – but they will hardly give up their resources. It is still the right sequence to first deal with the responsibilities. Once these are clear, you would outline that you need people to do the job.

## (ii) Technical teams doing data-related activities

Whenever you find out that a *technical* IT team has done a data job before, you can explain that the change makes sense. Interesting questions during such a dialogue could be

- Do they really understand your business problem?

- Are they solving it for you (or with you), or are they mainly providing the tools?

Based on the responses, you either have good arguments for the shift in responsibilities, or you find out that good data experts sit in a technical team and that they might be willing to move to the Data Office.

## (iii) Activities not yet covered

IT may have provided limited service, as it can often be observed in organizations without an overall data strategy.

Such incomplete service usually covers all technical aspects, whereas the critical activities of comprehensive management of data are missing, for example, the end-to-end view or the alignment with requirements across business functions.

Such cases provide you with excellent opportunities to explain the idea of business-driven data management.

# Correct data handling jeopardizes my project

People are often afraid that if they follow the new data rules, their initiatives will become too complicated, will finish late, and/or run out of budget.

It is worthwhile digging a bit deeper here.

A project manager or product owner is usually not against doing it "right first time." They just don't want to look bad.

Give them good arguments.

- Either: We can do far better for the entire enterprise if we add a month to get the data aspect sustainable.

- Or: We won't get the green light from the Data Office to go live if we don't do it right now.

You can support the last point through a system of data-related transparency:

- Every project needs to be data reviewed, and the results need to be made public.

- Criteria for temporary concessions need to be defined.

- All of this must get endorsed through the data decision bodies you need to put in place.

All in all, you don't make life difficult for people like project managers or product owners. Instead, you provide them with good arguments for additional resources or budget. And hardly anybody in charge would refuse the opportunity to get enough funds and people to create a sustainable solution.

## What if you fail?

People don't know whether a Data Office is going to be successful and whether it has come to stay.

Particularly, politically conscious people will be reluctant to openly support the new Chief Data Officer as long as dedicated Data Management could turn out to become a failed experiment. They don't want to back the wrong horse.

Let's be honest: Unless you have already covered yourself with glory in your organization before (or you are having dinner with the CEO regularly), those concerns are justified.

It helps in such a case to look beyond your own organization. While dedicated management of data may indeed be unchartered territory within your organization, it has become an established approach outside, across all industries.

And the increase in Chief Data Officers is not only steep, but it has also been lasting for a while – too long for a temporary phenomenon.

No matter how confident you are that *you* have the right recipes for your organization's handling of data, it is wiser to stress that Data Management hasn't been invented by you.

Many concrete examples are available from other organizations, beyond those dealing with data as their main subject. Your message, backed by statements from the big consultancy firms, could be the following:

*We are not the first organization with a Data Office. Others have done it successfully before. And if we don't do it, we will fail, or we will be overtaken by our competitors. We simply cannot afford to ever go back.*

In parallel, you can encourage individuals to work with you, by promising that you as the CDO will take the risk – and the blame in the improbable event of a failure.

# It has worked well without a Data Office

I still hear that corporate veteran say "We have been successful for decades without a Chief Data Officer!"

I had to admit that he was right. And nothing had happened within his organization to make a change necessary.

Well, not *within* the organization…

Ask people to look outside.

First of all, things develop. Technology develops, behavior develops, and expectations develop.

Secondly, the speed of development increases. While a chosen setup worked for decades during the last century, disruption has become the norm today.

And which area was impacted most heavily?

No, it was not technology. Despite a tremendous progress in this area, many base business models have remained unchanged. Cars, trains and planes still bring humans from A to B, computers still have a screen and a keyboard, and movies keep being shown in a rectangular format.

We should expect fundamental changes to come in all of these areas during the coming decades. For now, however, organizations need to react to the changes that have already started to materialize.

Yes, I am talking about the world of data.

Of course, we will see a lot of changes in this world in the future. But unlike many other areas, these changes are already in full swing. The exploitation of data left the laboratories years ago.

The future of data has already begun.

## I don't want to change…

Why has Change Management as a discipline gained so much attention during the past couple of years?

While the world around us is changing at an increasing pace, the human ability to adjust to change hasn't changed much. Change is perceived more as a risk than as an opportunity. People rate the impact of negative consequences higher than the value of possible wins, without necessarily being aware of this bias.

Starting to actively manage data means a change. That is why Data Management means Change Management.

The fear of the unknown doesn't come as single statements that you can respond to with a prepared answer, as it is possible with other, more fact-based kinds of concerns.

Instead, you would want to systematically address the fear. Some thoughts:

(i) **Be a shining example.**

Don't ask others to embrace change, but tell others how *you* do so. Others will watch you. They should see that you welcome change even if it puts your safe position at risk.

And, of course, don't just act as if you did. You will really need to be open to change, including possible future challenges to your own current way of managing data.

(ii) **Create a positive atmosphere.**

Data can be an exciting thing. You can demonstrate opportunities. The target could be to make people curious.

Give space for the expression of concerns. Let people share their questions and their ideas, and deal with all of them transparently.

You can market the Data Office similarly to a medical advisor – who helps you with your current problems *and* teaches you how to take care of yourself on your own from now on.

(iii) **Address the fear of the unknown.**

Many colleagues might be afraid of not being able to cope with the new way of working. After all, Artificial Intelligence and Machine Learning are synonyms for "magic" to many.

Your message can be "You don't need to understand all of this in detail. Only a handful of people need to be true experts. Data is here to make life easier for the rest of the organization."

# Will an algorithm replace me

People will be afraid that decades of experience and expertise are rendered useless.

And, we need to be honest, a certain number of specific qualifications will lose relevance in this context.

People with skills that will not be required at all may be afraid and rightfully so. But keeping the status quo will not solve the issue: Other organizations will gain a cost advantage, which puts even more own jobs at risk.

I consider two messages critical at this point:

(i) **Upskilling is possible.**

Hardly anybody will become redundant beyond recovery. It is trained skills that become superfluous, not talent or intelligence. Training can make use of the latter to develop future-proof skills.

(ii) **AI and humans complement each other.**

AI and humans have complementing roles – where each of the two can achieve 90 percent, the combination of both may be able to reach 97 percent.

Furthermore, AI is still focusing on repetitive, less complex tasks. This may allow for humans to concentrate on more challenging tasks – which are usually the tasks people enjoy more!

Take the analogy of a possible tennis hybrid:

- Humans with their creativity will still need to decide which style beat and where to shoot the ball, based on the immediate situation, the impression of the opponent's condition, and including the surprise factor.

- Machines with their precision could ensure the ball hits the ground exactly where the human player wanted it to land.

Another aspect in favor of humans is the need for quality assurance: For the time being, you would not want AI to validate AI products. You still need the human factor here.

Consider the upcoming species of AI solutions used to discover cancer from skin images. How are these solutions sold to the doctors of this world? Instead of hearing "AI will replace the medicals," they receive the message: "This is a tool that makes you stronger."

You might wish to prepare a similar message for your workforce. Not to keep people calm – but because it is true.

# Psychology of Governance

*"Fake news!"*

**Figure 15-1.** The data may well be on your side – but that might not be enough!

© Martin Treder 2020
M. Treder, *The Chief Data Officer Management Handbook*,
https://doi.org/10.1007/978-1-4842-6115-6_15

# Don't claim covered ground

Proper data governance is indispensable. But does it need to be "governance invented by the CDO"?

People will have difficulties accepting a policy that replaces a previous one that has seemingly worked well.

So why should you want to change such a policy?

If you incorporate existing policies into your overall data governance framework, you can win the original authors as allies, as well as those who are happy applying those original rules.

Another reason to consider the availability of a policy or standard "one burning issue less" is that it allows you to focus on the remaining, true gaps.

You can claim ownership of those data governance elements without having to change them. And where adaptations are required, for example, to consider new decision bodies, you can stress that the core rules remain unchanged.

Sometimes, you can even generalize a rule or a standard that was initially developed for a single business area. You can usually expect someone whose work result gets applied beyond the own remit to be very loyal.

This is less about finding the best solution by all means, but to win people through empathy. If there is any room for improvement with an existing policy or rule, you can still develop it further over time – ideally together with the original authors.

# Design an acceptable starting setup

It is wise to shape a generally accepted structure in which Data Management has its clearly defined role.

- A model could be: All foundational work is taken care of either by IT (databases, infrastructure, licenses, etc.) or by Data Management (data rules, governance, business alignment, glossary, etc.).

- You might accept (for the time being) that other teams are responsible for areas you'd actually consider part of a Data Management Office. This may, in fact, help you get buy-in and mitigate the inevitable feeling of competition for power.

- If you do a good job addressing the pain points, the organization (usually the Board) may consider asking you to take responsibility for the other areas as well. But don't push for it yourself unless you have a good reason to do so.

# Base your authority on accepted authorities

A general mandate from the Board is a must, but it may not help you in your daily work, and you might not want to hold it up like a police badge all day.

It is therefore helpful to check what else could strengthen your authority – even if it may not do so formally.

You should be on the safe side if you translate known organization strategies into data strategies: "To achieve (organization strategy X), we need to focus on (data strategy Y)."

Beyond this, you could try to quote Board members and other influential executives. You may do this as side remarks, always avoiding to sound arrogant.

Those that actively support you may even confirm that you are allowed to quote them ("Our CFO asked me to also consider...").

But you may also quote those that say or write something in support of your story. If you listen actively, you will hear many statements you may find useful. They may, for instance, state that they consider data valuable, or they may stress the need to focus more on data before taking a decision.

Whenever you take the minutes in a meeting (which is a powerful tool to frame agreements!), you would always document any such statement, and you may even put the required words into the relevant executives' mouths by asking the right questions.

Ultimately, people should gain the impression that a vast majority of decision-makers are supporting the direction of the CDO and the Data Office.

# Balancing two extremes

If the answer to a structural problem were the selection of the right extreme, we'd have far more successful leaders: You'd have a 50:50 chance of getting it right, without any expertise.

Instead, successful management means finding the right balance between the extremes. Such an approach consists of an initial thinking process that has a lot to do with attitude, followed by a permanent adjustment process, based on close observation and active listening.

There are various areas of balancing. Let me describe four of them as examples.

## Between absolutism and democracy

Absolutism would mean "I know the answer. I set up the rules. You need to follow." Many departments are tempted to follow this approach: We are the specialists, and the others are not. Why involve them in the development of rules?

Democracy would mean "Don't set up a rule unless there is a majority for it." With this approach, two things might go wrong.

Firstly, you might face situations where all departments would have to give in for the greater good (i.e., the shareholder view), and democracy would lead to majority votes against the benefit of the shareholder. The result is either a silo mentality as all parties agree to "do their own thing" or a financially unhealthy decision that a majority of stakeholders benefit from while the overall organization does not.

Secondly, important yet painful decisions (e.g. financial discipline or legal compliance) might get blocked because most departments don't like them.

But there is a position somewhere between the two extremes, both from an overall policy perspective and in dealing with concrete cases:

(i) **Governance**

Democracy can work if the room for decisions is limited in a way that it prevents decisions with a negative impact on the organization – just as a constitution prevents a majority from, say, suppressing a minority in a democracy.

First of all, a framework of guidelines helps. It needs to be sufficiently generic to be dealt with at the Board level (e.g., "Safety above all" or "Compliance above all").

Furthermore, an endorsed data strategy, as well as a list of Data Principles, would allow stakeholders to use their management freedom without endangering data targets.

(ii) **Case by case**

Taking a decision with authority does not mean ignoring other opinions.

A practical approach is, therefore, to first involve all stakeholders, listen to all arguments, and then devise a solution.

A final review of that solution by all stakeholders makes sense as a sanitary check. It would allow for disagreement only in conjunction with good arguments.

If you finally decide (or submit your solution for data council approval), nobody will be taken by surprise, and everybody had the chance to provide qualified feedback.

# Between centralized and local solutions

Strictly central approaches fail in considering local peculiarities.

Local approaches, however, tend to create locally optimized solutions. Slightly different local solutions may already lead to different "best solutions," resulting in a plethora of coexisting solutions.

Again, you will find the best solution through a balanced approach.

You may strive for a centralized solution but involve all local stakeholders and your virtual network of functional experts.

Their input needs to be substantiated, and it needs to be looked at in the light of an overall business case, not only a local one.

It will be critical in this context to have the overall business case to consider the "costs of complexity," that is, the impact of having to maintain multiple different (often local) solutions or solution aspects.

This impact goes up exponentially, particularly if you consider the risk of errors and the time to market of future fundamental changes to the entire ecosystem.

# Between standardized and individualized

Every CDO faces this dilemma earlier or later: Data Architects care about structure, and Analytics folks care about their freedom.

Similarly, you have database designers who want to fix the schema, while business folks want the freedom to have applications and databases reflect any changes in business logic quickly.

Thank goodness, you don't need to decide between these two extremes. An excellent approach between "standardized" and "individualized" is **"configurable."**

This approach has an impact on your corporate data model: If possible, valid scenarios should not exist outside the data model, not even as "approved exceptions". Instead, try to consider all *valid variances* as part of your data model.

The advantages are that you force business folks to think about their business targets and preferences while at the same time documenting the entire business model in a way that software architects can implement the data structure without ambiguity.

Of course, this comes at the price of a more complex data model. But this price should not be too high. Any additional level of detail helps avoid misunderstandings when Architects and software developers cast the business logic in applications and database structures. And if people consider the resulting data model too complex, remind them of the fact that this data model reflects the organization's business model – which may, therefore, be too complex, as well.

## Between dirty and perfect

Let us define the effort for the ideal solution to be 100 percent. How much of it does it cost to cover the happy flow, without any exceptions or deviations?

A mere 5 percent of the entire effort, at most.

People under pressure tend to aim for precisely this: Leaving out proper exception handling, thus reducing costs and time.

A primary purpose of data-based solutions is to address deviations from the happy flow. The reason is that such deviations make up for the lion's share of business costs.

That is why nobody should stop at 5 percent!

But do you ever need a 100 percent coverage? Probably not. The famous 80/20 rule often applies to data as well.

Covering 80 percent of the perfect solution with 20 percent of the effort is far better than covering nothing but the happy flow with 5 percent.

And, remember, you don't need to give up on the remaining 20 percent for good. Register all gaps as Data Concessions, and you may have them closed over time.

## Shape your data brand

"Data Office" may sound very technical to many employees. You should turn it into a brand that people can relate to.

If your organization policies allow, think of creating a logo – simple, colorful and positive. And use that logo whenever you deliver a service or publish a message.

It is also a good idea to simplify the job of the data office so that people understand it more easily, as illustrated in Figure 15-2.

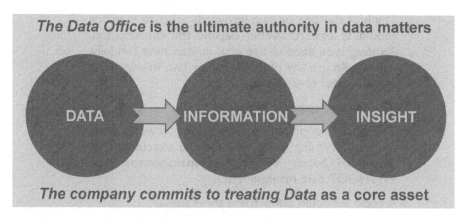

**Figure 15-2.** The Data Office in short

# Elevator pitch

Everybody knows how important an elevator pitch is. Is yours up to date?

There is not *one* elevator pitch for data management, of course. The story depends strongly on the situation of the organization, as it depends on your own ideas. But some patterns are worthwhile being followed.

Try to think of a common statement (your key message, independently of the target audience), followed by a target-specific message. Remember: Everybody sees different problems and judges opportunities differently, often based on their own roles.

(i) **The universal message**

Think of a one-sentence storyline that burns into people's brains. Something like "I think we need a dedicated data authority to stabilize operations and to allow for fact-based decisions."

Here we face a specific challenge with "data": Many people, including most executives, don't know what it is, or they think you talk IT.

This situation may require an intro – not more than one sentence either, though. Sometimes executives don't need to understand the entire concept – they just need to believe you. Quoting external, generally accepted authorities may help.

Your target should be to make your counterpart ask for more information. You can either make a bold statement or simply make the executive curious.

Think of the one or two keywords a Board member should at least remember beyond the conversation. Although you need to use your limited time carefully, you should use any of these one or two keywords as often as your time allows.

(ii) **The target-specific message**

To prepare a good (and short!) story, you would want to list everything meaningful for the executive that would be better WITH data management than WITHOUT data management.

A good list could cover these three parts:

- Where are we not reaching our targets today (but data management could help)?

- Where could we even exceed our targets today (through data management)?

- What are (data management) opportunities for a better future?

Now you can pick the #1 point of each of the three topics. This makes three good arguments, spanning a wide business area.

Assuming that data is not in focus for decision-making at all, a typical elevator pitch in preparation of a Data Management Office could be:

*To which extent do you think we base our decisions on data today? Do you always feel confident?*

*Did you know that we have (or could acquire) a lot of data that could help us take better decisions?*

*Our organization has a lot of data outside your area that could be of value for you.*

Or, if the organization's main issue is that each function does its own thing:

*We could gain far more insight from our data than we do today. All functions do something on their own. This results in a lot of duplication and ambiguity. And no function has the full picture.*

Where executives want the fancy stuff and don't see the full data picture:

*To be honest, it is not sufficient to simply set up an analytics team. A lot of groundwork is required, bringing together all business functions. Data may be wrong or incomplete. That is why data needs to be managed wherever it is touched – which is basically everywhere.*

Try to get the foot in the door – a verbal commitment that you can refer to when following up later:

*Will you support me when I present my concept to the Board? Would you have 30 minutes for me to share the concept with you up front?*

(iii) **If a Board member asks about your proposal**

*I am thinking of a dedicated, empowered Data Office. Data should not be a side activity of an existing department, not even of IT. To operate successfully, I am thinking of two steps:*

*Step 1: I would like to obtain an executive mandate, as my success should not depend entirely on voluntary acceptance.*

*Step 2: It is then up to me to convince all functions to engage. After all, I want to support them, not to take anything away from them.*

As with most explanations, less is more. If something remains unclear, the Board member will ask.

# Practical Aspects of Data Management

# Data Business Cases

**Figure 16-1.** The ROI dilemma

© Martin Treder 2020
M. Treder, *The Chief Data Officer Management Handbook*,
https://doi.org/10.1007/978-1-4842-6115-6_16

# Business cases for data – why?

Think of the typical Hype Cycle that organizations like Gartner work with.

We are coming from a time during which particularly the most modern data disciplines around Artificial Intelligence (AI) and Robotic Process Automation (RPA) were expected to replace human intelligence soon.

Each organization is setting up departments for AI, Machine Learning, or Data Science so as not to miss the train. This is what Gartner calls the "Peak of Inflated Expectations."

Meanwhile, we got beyond the point of "We need to do it unconditionally." This does not mean that topics like AI and RPA have failed. They haven't, and everybody knows of organizations that have become successful by deploying such data disciplines.

However, we have reached a point where leaders expect data to increase the profitability of their own organizations. This is why we hear "Show me the money" more frequently when asking for data projects to get funded.

And rightfully so! Data disciplines that don't contribute to an organization's targets, be it directly or indirectly, will have to mature at universities (to exaggerate a bit).

Have we already arrived at Gartner's "Trough of Disillusionment"? Not yet, but we need to work hard to keep it shallow and to enter the "Slope of Enlightenment."

One key contributor: The business case. You should follow the principle "No project without a case!"

Remember, this is not about the technology behind exploring data. It is about the confidence of the investors and budget holders in the ROI of data.

For this purpose, business cases are the right choice. Whenever a decision for an activity is to be taken, a business case is supposed to help base it on facts.

*Using facts is an excellent way of justifying data projects* – you can base on figures what people don't understand intuitively.

On the other side, business cases require numbers – not only about costs but also about expected benefits.

This is a challenge particularly for data projects. Many of them don't deliver direct value but enable other activities to provide such value.

This ambivalent situation justifies a dedicated chapter of this book. Let's have a closer look at business cases from a CDO perspective.

# Business cases in a perfect world

Before diving into solutions, we should first define what a business case is, then understand the challenges and opportunities that come with it.

## The fundamental idea behind a business case

**Step 1**: Compare the investment in a project

- With the **best possible alternative** investment (if the money is available) and/or
- With your organization's **cost of capital** (if the money needs to be borrowed)[1]

**Step 2**: If the project delivers a higher return, run it!

This is admittedly a very simplistic description – but the reality of business case handling in many organizations is even more simplistic.

## Capital cost over time

Typical calculations consider capital cost over time.

- You want to know whether the project's return is higher than the interest rate of the (assumed) best possible alternative investment, the **Internal Rate of Return (IRR)**.
- First, you determine all money spent or earned through your suggested project, together with the "Effective Date" (i.e., the day where the money comes in or goes out).
- In a second step, you calculate all positions toward one reference date (usually "today") by applying the interest rate backward or forward. If you calculate back, for example, from the future to today, this is called "discounting" (see Figure 16-2).

---

[1]Companies usually maintain such a standard interest rate

**Figure 16-2.** Discounting future values

- If the reference date is "today," the result is called the **Net Present Value (NPV)**.

Admittedly, this includes a lot of unknowns – but it is better than gut feeling alone. Note: This also applies to data management in general.

---

### EXAMPLE I

Let's assume **$95** are due today and you will receive **$100** in one year from now.

Let's assume an IRR of **5 percent**.

The business case will look at today's value of the two amounts:

1) The $95 which are due today.

2) The amount that will have become $100 in a year, using the IRR of 5 percent: **$100 / 1.05 = $95.24**

   (Formula: future amount / $(1 + IRR)^{years}$)

The overall NPV is $95 – $95.24 = – $0.24 (negative!)

The organization would earn money – but lose money compared to a predefined, alternative investment!

---

In such a case, not doing the project is the better choice for the organization as they would earn more if they invested the money elsewhere (or don't need to take a loan) – provided their IRR figure is accurate.

## Consideration of "risk"

Money expected to be earned in the future has to be considered "at risk."

- Meanwhile, alternative solutions may enter the market.

- Or the targeted solution, for example, a product or piece of software, has a limited planned life cycle so that benefits diminish.

Of course, you can devise complex formulas to determine a time-dependent "risk factor." However, as you will know, the future comes with uncertainty.

Figure 16-3 illustrates that uncertainty often makes it impossible to even calculate the break-even point.

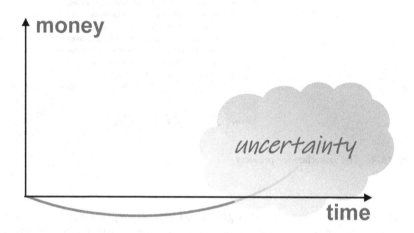

**Figure 16-3.** Uncertainty and time

That's why most organizations simply cap the return calculation, for example, after three years. The business case disregards any benefits that come later.

---

■ **Note**   Calculating accurately is one thing – whether all parameters stay the same is a different story.

---

## Project selection

If many projects have a sufficiently high IRR, an organization may cap the number of projects to stay within a predefined budget for projects. As a consequence, projects can remain below the "waterline" even if they are calculated as "worth being run." They will not happen.

**Projects**

- Lorem ipsum dolor sit amet, consetetur
- sadipscing elitr, sed diam nonumy eirmod
- tempor invidunt ut labore et dolore magna
- aliquyam erat, sed diam voluptua. At vero
- eos et accusam et justo duo dolores et ea
- rebum. Stet clita kasd gubergren, no sea
- takimata sanctus est Lorem ipsum dolor sit
- amet. Lorem ipsum dolor sit amet, consetetur
- sadipscing elitr, sed diam nonumy eirmod

"waterline"
- tempor invidunt ut labore et dolore magna
- aliquyam erat, sed diam voluptua. At vero eos

**Figure 16-4.** The waterline for projects

Alternatively, an organization can get additional money, for example, from a bank or the capital market. This makes sense as long as the applicable interest rate is below the project's IRR.

In all cases, the availability of resources has to be taken into consideration as well. Specialist knowledge may be a scarce resource that you cannot simply acquire on demand. This is another cap that limits the number of projects.

## All good?

This sounds like a scientific, unbribable approach. It will help get all data-related projects funded and approved as long as they deserve to be run from a shareholder perspective.

Or not?

Let's have a look at some pitfalls that come with business cases before we look at options to play the business case card in your favor.

The sad truth is

- Business cases come with several challenges.
- **Data** business cases come with a few *additional* ones.

Let's go through **nine** of these challenges: **Five** general ones followed by **four** that are rather typical of data projects.

# General challenges

## Business case culture

Let's do a quick quiz: Why are most people *really* creating business cases?

- To find out whether a project should be approved
- As a lever to get an initiative approved by all means

You guessed it: Unfortunately, it is NOT the first choice…

A project manager's typical attitude, as often encouraged by immature organization culture, is the following: "I have to 'tune' it until it is 'good' enough. If it is not good enough to get the project approved, they will consider me a failure. And I need to get some projects approved so that I can show my value." Sounds familiar?

This behavioral pattern can be observed in many, if not most of the larger international corporations.

It creates a suboptimal allocation of funds, and a lot of energy is wasted on "making the case look good."

Here is a rule of thumb for an expedient business case culture:

---

**DON'T REWARD POSITIVE BUSINESS CASES.
REWARD HONEST BUSINESS CASES.**

---

## Quantification of benefits

Can all drivers of a project be quantified?

How about these three aspects?

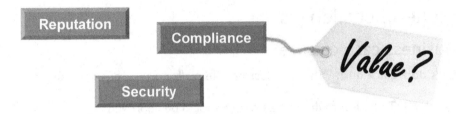

**Figure 16-5.** Values that are difficult to quantify

Try to quantify the impact of a bad reputation on customer loyalty. This would be truly important when justifying compliance work.

## Conflicting targets within the organization

Goals can be in conflict with each other, even if every single one is valid. A typical set of contradicting objectives is that of **sustainability**, **agility**, and **efficiency** – each of which is truly desirable on its own – see Figure 16-6.

In Operations Research, such a situation is usually addressed through a target function where the optimal weight of the different targets is the subject of the optimization task.

However, in real life, the parameters a, b, and c of the target function

**a * sustainability + b * agility + c * efficiency → max**

cannot be determined objectively, and every business function has a different perspective. While the rewarding scheme for managers should aim at aligning their personal objectives with the overall organization targets, there is always a gap.

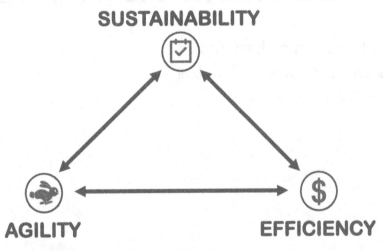

**Figure 16-6.** Balancing sustainability, agility, and efficiency

Each of these three targets has a value which needs to go into the business case. Risk Management would even add the *probability* of events – another aspect you cannot determine unambiguously.

For data projects, dealing with all of these aspects is a task of the Data Office. A sufficiently mandated CDO should be in a position to represent the shareholder view better than other executives, based on the cross-functional, long-term perspective that is an essential part of a CDO's role.

The biggest challenge of the CDO is the adequate support of the sustainability target – experience shows that very few stakeholders within an organization's management support it actively. Sometimes, it's only the CDO and the CEO. If the CEO is about to retire or move on, it might be the CDO alone.

Let's have a look at other typical players in an organization:

- A typical **project manager** aims at maximizing agility so that the project can be completed quickly, without too much "bureaucratic overhead." Very often, project managers are not rewarded for sustainability.

- The **CFO** focuses on efficiency, that is, the maximum value for money, ideally within the fiscal year. This includes a look at the marginal costs (MC): As soon as an additional amount spent delivers a return below a certain threshold, the CFO's willingness to provide the money will reduce dramatically.

- **Long-term investors** and owners usually put sustainability high up on the list. For them, it is okay if it takes a bit longer, as long as no future costs are incurred.

- The **Legal department** will insist on compliance – which is part of the sustainability target. To this department, the other two targets may be secondary.

All of these players will need to get involved, but you will need to force them to quantify their preferences – which will allow for an overall cost/benefit comparison from the organization's perspective.

# Difficult validation in retrospect

Imagine the typical process of creating a business case:

- You diligently calculate the costs and benefits for the business case.

- Then you run the project, and you complete it successfully!

- You know exactly how much money you have spent.
- But ... *can you prove the benefits???*

Issues:

- Claimed benefits cannot be validated easily.
- Determined benefits cannot be unambiguously allocated to a single project.

You should foresee the validation of a business case to be in a better position for your next business cases. If you are not convinced of your own business case, you shouldn't push for the project anyway.

**Figure 16-7.** Benefits that are difficult to quantify

## Quickly outdating business cases

Imagine you calculate the value that your project will generate. And it is accurate and precise.

But times change...

In fact, the further an expected return lies in the future, the more uncertain it is.

For example, new technical development may render a solution useless.

# Data-specific challenges

Business cases for data initiatives sometimes don't come without irony: You need a business case for your data, and you need data for your business case.

*"Yes, Sir, I tried to build an ROI case for our BI project -- but I couldn't access any reliable data!"*

TimoElliott.com

**Figure 16-8.** Business case Catch-22

This cartoon reminds us of the fact that some challenges are observed more frequently with data-related business cases than with the average business case.

But what do we refer to as "data-related business case"? It is short for business cases for initiatives aiming at capitalizing on data or data handling. These can be as diverse as

- Introduction of cross-functional MDM
- Subscription to external data sources
- An organization-wide hackathon
- Creation of additional Data Science positions
- Prescriptive Analytics based on the behavior of website visitors
- Migrating from Hadoop to Spark (or any other IT project about technical data handling)

Here are a few of those challenges most frequently found with business cases for data initiatives.

## Data is not sexy

Here's another quiz. What's wrong with the following story?

---

### EXAMPLE 2

During the evening get-together after the annual executive conference, the Head of Machine Learning is having a chat with the Chief Commercial Officer, a lady in her 30s. When we get closer, we hear him say to her:

"…and then we improved the Neural Network by giving up using a simple Hard Sigmoid function as an activation function in favor of the Gudermannian function

$$f(x) = gd(x) = \int_0^x \frac{1}{\cosh t} \, dt = 2 \arctan\left( \tanh\left( \frac{x}{2} \right) \right).$$

This requires more computing power, but we observed far better results!"

The Chief Commercial Officer replies "Cool! I cannot wait to let you present the technical details to the Management Board!

Once they understand how it works, they will immediately provide the budget for the additional hardware!"

At the same time, she looks at the handsome Machine Learning guy and thinks "He is sooo cute!"

---

Have you marked all the errors?

Here's the solving:

a) You won't see a Machine Learning person invited to an executive conference.

b) No executive will listen to a nerd.

c) A Board member wants to know WHETHER it works, not HOW it works.

d) Don't hope for any budget because the Board members understand "data." They don't.

e) Data knowledge doesn't make you sexy! (Sorry for having to be honest here.)

# Determining the benefits of enablers

Data projects are often "enablers": Their value-add comes through future activities. Examples:

- MDM improves the stability of operations.

- Analytics gives Marketing the required market insight, avoiding money wasted on futile marketing campaigns.

- SOA[2] avoids data inconsistencies, thus reducing the number of manual corrections in Finance.

- Data Science discovers previously unknown correlations, allowing for geography-specific service configuration.

Why do enablers often remain unfunded, even in the absence of political or cultural issues?

You frequently face the following two issues:

- Little to zero *direct* financial benefit.

- Future projects claim the benefits.

As an analogy, let's have a look at an enabler project outside the data world.

| EXAMPLE 3 |
|---|

The "Enabler" of Burj Khalifa: Its foundation!

Over 45,000 $m^3$ of concrete, weighing more than 110,000 tonnes, were used to construct the concrete and steel foundation of this tower, which features 192 piles buried more than 50 m deep: 20 percent of the tower's concrete is hidden below ground level!

What would the business case for this tower's foundation have looked like in isolation?

*Tremendous costs, no single room, just a platform…*

This case already indicates a first possible approach: The foundation was not approved and funded in isolation – but together with the tower on top, as one single project.

---

[2]Service-Oriented Architecture

## Late break-even

How quickly does an investment have to break even?

A project should be run if it "makes money" in the long run.

In reality, an organization will get as many projects funded as it has budgeted for. All other projects will not happen, even if they promised a net gain to the shareholders.

Which projects will be the first victims? It is often not those with the worst ROI but **those with an ROI in the far future!**

Figure 16-9 indicates a case where break-even is only reached **after** an organization's defined reference period. Such an organization will calculate a negative ROI for such a project.

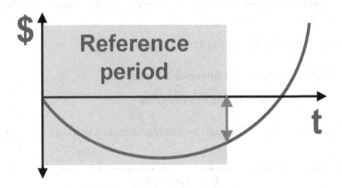

**Figure 16-9.** Break-even after the reference period

Not surprisingly, a lot of foundational data work falls into this category.

Here are some helpful analogies to explain the situation of data projects to an executive:

a) Anything other than a long-term view would prevent an airline from buying new planes.

b) In forestry, NOT to plant new trees when cutting the grown ones saves money. This approach, however, will kill the organization in the long run.

---

■ **Note** As soon as you treat data as an asset, a long-term focus would even increase an organization's valuation today! (See "discounting" earlier in this chapter, section Capital cost over time)

---

## Motives of business leaders

As you may have witnessed, many projects get rejected for nonfactual reasons.

So, why do leaders not support all of the positive business cases?

Yes, sometimes they *really* don't understand.

In many cases, however, they understand but

> Don't see a benefit for themselves
>
> Are afraid of the change
>
> Rather follow their own agenda

The particular challenge with data business cases is that business leaders get away with it more easily! None of the other executives knows better, and hardly any of them would defend or support a business case of an initiative that they are wary of.

# Eight secrets of data business cases

Instead of complaining about something we cannot change, we should look at how to deal with the situation.

Looking at all of the possible obstacles to get your business case approved, you need a versatile strategy!

While this strategy will look different from organization to organization, I'd like to share my eight secrets of data business cases – which can form part of your strategy.

## I. Do active stakeholder management

You remember the last challenge of data business cases: Lack of executive support for what you perceive as obvious must-dos.

This is why stakeholder management is key. Understand them, create trust, shape win-win situations, and be ready to compromise to a certain extent.

How do you do this? Get in contact. Have individual meetings – formally or over a coffee.

The key aspects are

(i) **Ask and listen!**

This is my recurring message. It applies to business case management as well. You get to know the deciders' most significant pain points. And you can tailor your story accordingly.

(ii) **Make them curious! Cause shining eyes!**

Develop examples that are meaningful for a particular executive. The more concrete and tangible those examples, the better!

There is also an aspect of tedious work associated with stakeholder management: Documentation. Create your stakeholder map. Determine and note relationships between deciders. Confront deciders with the (positive) point of view of their friends or allies.

And you should classify your stakeholders. Here's a possible list of categories:

- The overchallenged: Interested but afraid

- The whiners: Always complained in the past but no contributors today

- The curious: Often clueless but ready to listen

- The self-contained: Think they can do it all on their own

- The superficial: Want the colorful stuff without the hard work

- The arrogant: Consider data management superfluous

- The committed: Your perfect partners

You need a plan for each behavioral pattern. And you can plan to move stakeholders from one pattern to another.

## II. Foster data literacy and transparency

People need to understand what you are talking about. You need to manage expectations and explain, to foster trust in data.

What do you need to do?

You need to teach the entire workforce on data.

Easier said than done?

Correct! But there are a few things you can do to improve the situation. Whoever understands data (or at least what data can achieve) will be more supportive.

First of all, be transparent, and be honest about all the pros and cons.

Now, it may be tempting to conceal risks to make your business case look better. But not doing so has a few advantages:

(i) **It helps enforce transparency.**

As soon as your cases turn out to be honest, you have the legitimacy to set the standard for all competing business cases.

(ii) **It improves your reputation.**

Managers who deal with stuff that people generally don't understand depend on gaining a reputation of being honest. After all, you cannot expect too many people to agree with you based on their scientific assessment of your plans. They need to trust you.

(iii) **It covers your back.**

Incorporating risks may make a project less attractive than an investment with the same ROI but lower risk. But it may save you in case of failure – you can refer to the risk.

That is why any risk should be incorporated into a business case.

Secondly, don't expect people to voluntarily "enter the land of data." You'd rather pick them up where they are. Use their verbiage. And think of analogies that are meaningful to people in their respective context.

A nice example is an increase in the value of data if you process it further. You can compare it with the increase of value from raw material like iron ore to a finished product like a plane, as illustrated in Figure 16-10.

This example will be understood easily, and it will allow you to compare it with the increase of value when you turn data into information and insight (e.g., by structuring it, cleansing it, bringing the right data together, etc.).

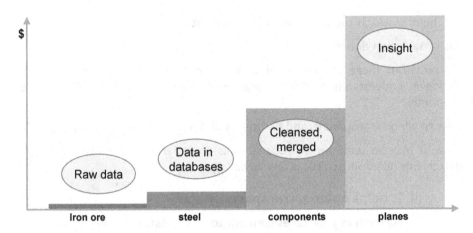

**Figure 16-10.** The evolution of value

In a shortened version, you could compare the frequent slogan "Data – Insight – Action" with "Steel – Planes – Air Transport." This comparison makes visible how complex the way from data to insight is, compared to the step from insight to action.

And you don't need to stop there. Just as planes only represent value once they bring passengers or goods from A to B, insight from data will only add value once the organization makes use of it, for example, by taking better strategic or operational decisions.

## III. Create and maintain a data road map

People are often open to catchy terms on a road map, as long as they don't need to commit to resources, funding, or compliance. That is why it may be better to decouple the two steps.

As a first step, you would describe the road map without asking for concrete resources or funding. You would generally explain why it makes sense to take a number of steps. You would then ask for everybody's commitment to your road map. People will probably agree if they don't find a good argument against your road map.

You can quote this commitment later when you detail out your plans into concrete projects and when you need funding and resources.

A typical reference to early commitment could be "This project is part of our approved strategy to modernize our MDM until the end of (year)."

Such an approach is obviously stronger if the commitment comes from the top floor. People may think twice before refusing to support a Board-level decision.

And this is not only about the mere funding process. It is also about making managers across the organization willing to provide resources to your initiative.

# IV. Treat data as an asset

Data is not only called an asset more and more frequently. It also has a lot in common with formally accepted assets:

- Using it brings tangible business benefits, for example:
  - We are more efficient with the asset than without it.
  - We can win new customers and keep existing ones.
- We can transform it into something with an even higher value.
- It degrades over time and requires maintenance.
- Having it and not using it is a waste.

In addition, intangible assets such as data even have advantages compared to tangible assets:

- They don't become less when used.
- They can be made available everywhere within seconds, and everybody can use them at the same time.
- While "sweating the asset" is often not sustainable when working with tangible assets, it is a perfect recipe to capitalize on intangible assets such as data.

Still, many organizations don't see data as an asset. According to Gartner's former VP Analyst, Data and Analytics, Valerie A. Logan, a mere "8% reported that key information assets are measured as if they were on the balance sheet" (Logan, 2019).

Or, as Thomas Bodé, Global Head of Data and Analytics at Daniel Swarovski Corporation, phrased it during a data conference in Barcelona in February 2020: "Data is the most undiscovered asset."

This situation should encourage you, as it indicates significant opportunities.

Even if regulations still prevent data from being treated as an asset, you should address the topic. The following activities may help you:

(i) **Attach a price tag to data.**

Data should obtain a price tag based on its potential value-add.

Whenever, as part of a project, the value of data (by "refining" it) increases, you should add this increase to the benefits of the project — even if the actual value-add lies outside the project. (But explain how you calculate!)

(ii) **Work closely with the users.**

Engage the business teams! You can jointly determine the added value of each step in the data supply chain, such as acquiring, cleansing, enriching, business metrics, or reports.

As a second step, get their commitment to the value-add of data.

A value tag that comes directly from the beneficiary is an entry ticket for business cases!

(iii) **Use analogies and examples.**

Sometimes a bold statement may help illustrate the equivalence, such as "Unmaintained customer data rots as quickly as forgotten apples."

In fact, with the increase in data turnover in our digital time, data could more and more be classified as FMCG (Fast-Moving Consumer Goods). Even if you are not in that industry, you might consider making use of this analogy.

(iv) **Ask the right questions.**

Help people. Don't simply ask "What's the value of this report?"

Instead, ask "What would be the impact of you not having it?"

(v) **Talk to the CFO.**

In most organizations, a promising point of contact is the CFO. This person is not only very influential but should also understand the character (and criticality) of assets better than anybody else.

(vi) **Establish policies.**

Even if there is no immediate impact of "data as an asset" on balance sheets, for example, as an "intangible asset" as per IFRS,[3] you may work on developing policies within your organization that allow for data to be dealt with similarly to other assets.

This will enable you to apply principles to data that are known to be acceptable for formally recognized assets.

*"I'm just off to the bank..."*

**Figure 16-11.** Data is an asset

# V. Work in stealth mode if necessary

Depending on an organization's maturity level of dealing with data, data-driven initiatives may be forced to run in "stealth mode" until first real-life evidence can be demonstrated.

---

[3]The International Accounting Standard (IAS) 38 describes "intangible assets." See IFRS (2017) for the official definition of the IFRS Foundation.

If the immediate business case is not showing sufficient benefits without considering future reuse, and if no one is interested in a long-term business case, it may be possible to "hide" the idea within another, approved project.

This would allow for the implementation of a few first examples. The value usually becomes evident with the next use case.

For example, there is no need to implement any data-related logic a second time – just call the web service (that was built for the first use case).

Have the project manager of the second case spread the word – and people will begin to believe in this concept.

You have reached your target as soon as people who would still like to implement it the dirty way would face resistance from others and subsequently decide to follow the rules.

## VI. Explain the cost of NOT doing it

As we saw, a solid data foundation often acts as an enabler, and all the benefits are attributed to the projects that benefit from this foundation.

Instead of attributing parts of those benefits to your data foundation as well (and risking being accused of double counting), you might accept the allocation of benefits to those implementation projects. The folks running those projects may be grateful for your support in reaching their projects' financial targets.

Instead, you might consider calculating the "costs of not creating a data foundation." This approach would allow you to list all projects that would either not have been possible or at least wouldn't have delivered as much return on investment without your data contribution.

## VII. Set up and follow business case rules

You will face situations where your business cases will be challenged. People may, for instance, claim that you are biased and that you have tried to make your business case look better than it should.

We have to admit that such accusation is often not pure invention. It becomes even more challenging to respond to where you work with indirect benefits, as it is often the case with data business cases.

It is, therefore, useful to be prepared, ideally through generally accepted principles for the creation of business cases. And people will more readily accept (justified) principles if they do not have to commit to budget and resources at the same time.

That is why you may wish to do it well before creating your first business case, when agreeing does not cost anybody anything!

Of course, you would include all the standard components as described earlier in this chapter, such as the "Net Present Value" calculation, or a maximum time horizon for the consideration of benefits.

But you can achieve more significant benefits for data-related projects if you manage to include soft factor rules as well.

One example is the "cost of not doing it" calculation as described earlier. This one should be high on your list of business case rules to become corporate standards.

A second example is the handling of *conditional* benefits. Just like a fire extinguisher that may never get used, you may never face compromised security or damaged reputation. But there are still good reasons for a thorough implementation of data protection mechanisms, just as there are good reasons to have a fire extinguisher.

A straightforward rule to calculate this kind of benefits is to multiply the impact of an event with its probability. As soon as there is an agreement on both probability and impact (positive or negative), the benefit calculation becomes easy.

# VIII. Develop a corporate data culture

Look around you: Does any function or department within your organization get almost all projects approved, independently of the people involved?

Why is that? Are projects of a particular type seen as a "matter of course," a natural "must-do"?

Depending on an organization's (explicit or implicit) culture, certain aspects are commonly agreed to be nonnegotiable.

- Maybe data protection is considered such an essential aspect that all related projects get funded.

- Perhaps there is one long-term transformation initiative with Board sponsorship so that projects always try to be considered part of it to obtain budget.

It should be your target to make "Becoming a Data-Driven Organization" one of those Board-sponsored targets or initiatives, maybe even an official part of your organization's culture.

# Use cases for data as an asset

Business cases are not only about financial justification. They are also about making a case to convince people.

## What drives your forecasts?

Here is a true story. I won't disclose the organization's name, and I have changed the numbers.

*The management of this organization knew that employees need to get motivated. They decided to foster a can-do attitude by sharing a tale from several years back:*

*After difficult times, the organization had to set targets for the years to come. All of the executives had proposed very moderate targets based on their pessimistic view of the world.*

*Frustrated by their unambitious targets, the CEO stood up and said: "We will go for a revenue increase of one billion dollar!" Everyone was shocked as this didn't seem realistic.*

*But, eventually, all of them accepted the target (actually, they had no choice). All executives felt challenged and started to work hard (which they would have done anyway).*

*A few years later, they had recovered from the crisis and reached their one billion dollar target.*

*Without an ambitious CEO, they say, these results would never have been achieved!*

*Years later, the management decided to use this very story to explain why they are setting ambitious goals. Skeptics are reminded of the success story back then.*

*Unfortunately, the economy had meanwhile weakened. The market was shrinking, and competitors were offering discounts to gain market share. Both revenue and earnings went southward, although the organization did well, both operationally and commercially. Ambitious targets had not helped. Instead, they had blurred the situation, as everybody thought hard work would make them achieve the target, as it had done last time.*

Here are three critical learnings for any kind of financial forecast:

a)   Ambition may be a useful ingredient of forecasting – but if you base your entire plan on a mixture of gut feeling and ambition, you could equally well use a random number generator.

b)   Ignoring external data makes you blind. Ambition does not protect you from market distortion.

c)   Mere extrapolation of past development is not a good idea either. The outside world is changing, and this cannot be derived from internal data.

How do YOU do your budget forecast – based on fiction, gut feeling, or data?

# Data – why NOW

It is important to stress that the basis for data business cases has changed tremendously during the past 10 to 20 years.

- According to Gartner,[4] by 2020, there will be more than 20 billion connected sensors, endpoints, and digital twins for potentially billions of things.

- The data available for analysis doubles every two years.

- Old algorithms can now be applied as the necessary computing power has become available.

- The development of AI algorithms progresses quickly, as it has become a focus area of mathematical research.

- The number of data scientists goes up. Neural networks are an increasingly attractive discipline as part of mathematics studies. Data Science is applied mathematics, after all.

- Customer expectations develop rapidly. For example, customers do not tolerate it if organizations ask for any of the information they have already provided. Customers know that this can be avoided.

# Consumer data

Can ordinary data work as an asset?

---
**EXAMPLE 4**
---

When I was young (I mean, *really* young), I earned a bit of money distributing advertising brochures from a tree farm.

Those brochures were aiming at people living in houses with gardens – other folks would hardly have any use for the trees sold.

I was sent to wealthier residential estates, and I was asked not to waste time or brochures on apartment buildings, on shops, or small firms.

Today, this job is done by the postman – who does not check whether an address is part of the target group.

---

[4]"Leading the IoT," 2017, edited by Mark Hung, Gartner Research VP, www.gartner.com/imagesrv/books/iot/iotEbook_digital.pdf

Without any local knowledge, the tree farm would, therefore, waste a lot of money sending brochures to families on the fourth floor who would wonder how to plant a tree on their balcony.

Now imagine you had data describing the character of each residential area: Is it a rural or urban area, is it a rich or poor settlement, and so on. That data could reduce the divergence loss significantly, ensuring you only distribute brochures to sufficiently wealthy rural areas.

One step further: Let's assume you have data that holds all addresses, including the kind of housing, the size of the garden, the value of the property, and other attributes. This would allow you to tailor your campaigns even better.

Finally, if you can combine this data with your customer file, you can determine to whom of your customers you should offer which product.

Given the benefits you'd associate with each level of sophistication, you can add a value tag to your data. Or, if you need to acquire that data first, you'd be able to determine how much you'd be willing to invest so that it pays off.

Over time, your data would age, and you'd have to refresh it regularly to prevent the inevitable waste from increasing.

If the tree farm had asked someone like young me to collect all of the necessary data while distributing the brochures, they would have gathered a significant amount of valuable data that they could even have sold to other organizations (only as far as GDPR allows, of course).

---

Now have another look at this example:[5] If the text had read "magic dust" instead of "data" – the story would have made sense as well (despite sounding a bit like a fairy tale).

The difference? Organizations could easily classify such "magic dust" as an asset. Shouldn't they at least *treat* "data" as an asset as well?

## DATA MANAGEMENT THEOREM #10

**Organizations should treat data as an asset, regardless of whether it is considered an asset from a legal or tax perspective.**

---

[5]Did you notice that this example is even GDPR compliant? There is no personal data involved, other than customer data with a justified usage.

# Data Ethics and Compliance

"*I'm just going to check if we have any dark data in the cellar...*"

©⓪ TimoElliott.com

**Figure 17-1.** There is a dark side to data

© Martin Treder 2020
M. Treder, *The Chief Data Officer Management Handbook*,
https://doi.org/10.1007/978-1-4842-6115-6_17

# Ethical behavior and data?

## What could go wrong?

Everybody knows that most countries don't allow for organizations to reject job applicants for their skin color or religious belief. The reasons are easy to understand.

But not all evil we could do with data is forbidden by law. A lot of it is simply too young to have been cast in paragraphs.

On the other side, some activities that are by no means meant to do any harm are forbidden, and organizations can get it wrong without any ill intent.

Taking all of this into consideration, how do you deal with data ethics, while being in charge of your organization's handling of data?

Let's first have a look at a few examples of controversial data technologies and opportunities:

- Optimizing on average while discriminating against individuals

- Accepting errors of algorithms at the cost of individuals

- Using personal data without the owner's consent to protect that person from irrelevant advertisements

- Using data against someone (as a weapon) to defend yourself

- Analyzing people to be able to influence their behavior

- Faking data or algorithms to change your audience's perception of a situation

You could think of some of these points to be outright wrong, while you might consider others acceptable, or you might say it depends on the circumstances.

In any case, the situation is not straightforward, and it is worth having a closer look.

## Where do we stand today?

As people begin to understand the opportunities, they become aware of the risks.

This includes the fear of the unknown: There may be no danger at all, but you cannot tell, due to lack of transparency.

In response, an increasing number of countries regulate the handling of data, and transparency becomes part of the regulation.

The United States focus on transparency following several accounting scandals. The Sarbanes–Oxley Act, forcing organizations to be transparent in financial matters, was already introduced in 2002.

Europe's primary focus is on data privacy. GDPR[1] went live in the European Union in May 2018, and after a de facto grace period of one year, penalties have started to hit organizations all over Europe. Non-European organizations are forced to comply as well if they want to do business in the EU.

Many colleagues in the United States and other non-European countries told me that they consider GDPR an exaggerated reaction to the world's data privacy challenges.

This, however, should not be a reason to be less compliant. I usually explain to my US-American friends that European authorities take GDPR as seriously as the US authorities do with SOX compliance. (Guess how I tell Europeans how seriously they need to take SOX.)

Countries on basically all continents have taken GDPR as a reference point for the development of own regulations, as the misuse of data increases with its technological possibilities, and it doesn't stop at country borders.

But countries have not waited for the EU regulations so that they can copy them. Examples of older laws are the "Cybersecurity Law of the People's Republic of China" from 2017 or the "Russian Federal Law on Personal Data (No. 152-FZ)" which even dates back to 2006, although extended in 2015 by the need to "localize" the personal data of Russian citizens.[2]

Still, a lot of potentially unethical behavior around data has not yet been regulated. But awareness is increasing among governments, and more regulation is expected to come, for example, in the area of face recognition. What is allowed today may become illegal tomorrow.

---

[1]The **General Data Protection Regulation** ("GDPR") "regulates the processing by an individual, a company or an organisation of personal data relating to individuals in the EU" (EU Commission, 2019).

[2]This is one of the laws that made it difficult for globally operating organizations to use standardized software and central data storage across the globe: Data captured on Russian citizens has to be registered and stored in Russia before a copy can be sent to another repository outside Russia.

It is interesting to observe that most regulations use a "consent-based model": Dealing with an individual's data requires that person's explicit consent. Yet, where individuals give their consent (and some monopolists may not leave them a choice), almost everything that is done with their data is usually legally allowed.

You may still wish to determine remaining restrictions. Many countries, for instance, do not allow for the localization of people under certain circumstances, even if personal consent was provided.

## Resulting questions for each business

In this situation, organizations need to ask themselves a few questions before defining their data ethics direction. Typical examples are

- Would we benefit from exploiting controversial data-related technologies and opportunities?

- If so, how sustainable are these benefits?

- Do we want to be ethical for the greater good or for economic reasons? Or a mix of both motives?

- To which extent does ethical behavior make sense for us?

- Does it pay off economically? Directly or indirectly?

- Does it even make sense to be ethical beyond what laws demand?

- Do we have a consensus within our organization about what is "ethical"?

## What are your options?

Organizations are not forced to choose between being ethical and not being ethical. Let me outline a few realistic options:

(i) **Case by case, based on business cases**

It is possible to decide for or against compliance based on a (possibly) complex business case. This would be done in each individual case where noncompliance is expected to come with some benefits.

The cost/benefit calculation would have to consider both the immediate financial impact and the indirect impact (reputation, etc.) of noncompliance.

Furthermore, the cost of getting caught would be multiplied by its expected probability.

Eventually, such an organization will accept noncompliance more easily if they think they get away with it.

For example, the decision to become GDPR compliant would become a business case of summing up expected penalties and the cost of adverse consumer reactions and of multiplying it with the probability of getting caught.

In essence, such an organization accepts to break the law if an increase in shareholder value can be expected on average.

You will not be surprised that such an approach comes with significant risks:

- Whatever impact you may calculate in case you get caught: You'd have to expect it to be worse than estimated. Judges will distinguish between organizations that have been too lazy to fix known problems and organizations that have consciously violated the law for profit reasons.

- Organizations can go out of business. The infamous company Cambridge Analytica has been tremendously successful: They did what they offered and what they got paid for. They were profitable. However, they violated privacy laws and had to close operations in 2018. Even if sacrificing this organization may have been a calculated part of the entire game, the story demonstrates the possible power of public reaction.

- Unethical behavior is not only jeopardizing the reputation of the organization. The reputation of management is at stake as well. Becoming associated with unethical behavior does not look great on your CV.

(ii) **Compliance as a nonnegotiable principle**

You decide to be law-abiding by all means.

Within these borders, the business case counts.

This approach sounds acceptable at first glance. At least you are on the safe side, as far as legal compliance is concerned.

The risks become apparent if you look at it from a different perspective: If it pays off, you basically decide to "do it unless it is forbidden."

So, what are the resulting risks?

Unfortunately, penalties are not the only negative consequences to be taken into consideration.

Everybody has observed the impact of shitstorms on social media. They make organizations look bad publicly, and they influence consumer behavior. And how many of the underlying activities were actually violating the law? Organizations are legally allowed to sell patented medication at prices that exclude huge groups of patients or to fuel their container vessels with dirty crude oil, to name just two examples.

If Cambridge Analytica had acquired all personal data with the owners' consent, public judgment would not have been substantially more favorable. The brand got destroyed primarily through its toxic business model.

(iii) **Ethics-based business cases**

Here you would again use a (potentially complex) business case, considering both short- and long-term costs.

But in addition, you'd acknowledge that there seems to be a market for ethical behavior.

You would ask: What are the risks of NOT considering ethical aspects while staying within the boundaries of the law?

In essence, you consider both the law and ethics in your business cases.

Are you on the safe side now?

I have to disappoint you here – there are still considerable risks left.

The world is more dynamic than ever before, and so is consumer behavior. Your business cases may not reflect customer behavior beyond today. It may be based on an outdated perception of customer behavior (based on figures from the past), or customer preferences may change soon after you have measured them.

When carbon-neutral service offerings came up several years ago, I recall an executive at an Express company ask "Who on earth would voluntarily pay more for a shipment just because it is transported in an ecologically friendly way?"

And, indeed, it was somewhat beyond imagination back then. Today, organizations insist on their shipments to be transported in a $CO_2$-neutral way, following the strong demand of their customers. No major transport provider could afford *not* to offer such a service.

But what do I recommend after having criticized all previous alternatives for dealing with ethics in the data space?

Here it is:

### (iv) Ethics as an organization principle

You might not wish to let the law be your only compass when it comes to acceptable handling of data, notably in the area of AI.

This is where your organization's culture comes into play. Without any guidance, a culture of "If we can do it, let's do it" or "It is okay if it is not forbidden" develops quickly. In such a case, it might be challenging to change employee behavior once regulation tightens or consumer behavior changes.

Remember: Many aspects of AI are too young to have been regulated so far. Expect further regulation to come soon.

If you get caught, the fact that something "used to be unregulated" is not a valid excuse. Regulators grant a transition period at best.

It is easier to create a culture of caution than having to monitor all activities of morally unguided employees.

## A broader perspective

We should keep this in mind: Ethically questionable AI is just an old behavior in a modern shape. Evil hasn't come with AI.

You could put it like this: There is a long tradition of ethically questionable HI (Human Intelligence). The discriminatory decisions taken by biased AI algorithms are comparable to traditional human discriminatory decisions. The main difference is that organizations can now blame the algorithm.

When Air Berlin went bankrupt in 2019, Lufthansa gained a temporary monopoly on some German domestic passenger flight relations. They immediately increased the fares.

When this became public, they faced a PR crisis, reaching from shitstorms to calls for a boycott.

Lufthansa hastened to explain that they use an algorithm to determine the best fares based on various factors such as demand and utilization. While this is not a lie, it was easy to see that the whole process is not an "act of God" that Lufthansa fell prey to.

And the algorithm did not take into account the damage of Lufthansa's reputation, which weighted higher than the relatively small gain through temporarily increased fares.

An organization culture based on ethics, together with a data-literate workforce, may have had all of this already considered when developing the pricing algorithms, for example, by incorporating an upper limit for the airfares.

Does all of this mean that we don't need to work with business cases for data anymore? You could look at it this way: The first filter is your organization's ethics culture, followed by the business case as a financial filter.

And how will your shareholders or owners think about this approach? Might they assume that you are trading shareholder value for ethics?

Luckily, you act ethically if you are selfish. Ethics increase an organization's value. A buyer's boycott destroys value.

Twenty years ago, hardly anybody would have assumed that one day a significant amount of people are willing to pay a premium for ethically produced goods or ethically provided services.

We know better today – let's have the courage to make our organizations successful by being ethical!

## GDPR – All done?

We have closed the GDPR project in time before May 25, 2018, and now we are done. Right?

GDPR forces us to treat data as we should have done anyhow. Eventually, the GDPR threat ensures we get support and funding to do so.

Many organizations have considered "GDPR" a one-off project that comes with an externally imposed deadline. And in many cases, the target was not to "achieve X" but to "come sufficiently close to being sufficiently compliant until the deadline."

While this is an integral part of GDPR compliance, I'd like to draw your attention to three other aspects of GDPR. In my view, these aspects are at least as relevant.

## You are not done after closing "the project"

Here is the first aspect: Don't stop because May 28 is over.

This may sound obvious as organizations will keep getting audited, and what an organization may have gotten forgiven early is expected by the authorities now. They assume that, meahwhile, all organizations have had enough time to become genuinely compliant.

But reality shows that attention and funding quickly go elsewhere. GDPR disappears from leaders' radar screens, and new, more interesting topics come up quickly.

## Privacy must become a way of thinking

The second aspect is about GDPR as a way of working instead of a one-off project that got launched, executed and hopefully closed more or less on time.

GDPR is often perceived as "coming on top of people's day jobs." Instead, it should become *part* of their day jobs.

Privacy needs to become a principle. The first "project" should be about establishing data privacy in our targets, not about implementation.

"*Sir, some citizens have been complaining about our data privacy policies...*

*Do you want their names? Social security numbers? Most embarrassing secrets?*"

timoelliott.com

**Figure 17-2.** Data privacy at its worst

# See business opportunities

My third aspect is a genuinely positive one: GDPR does not merely make life difficult for an organization. It also allows for new business models, and it will provide opportunities to differentiate from your competitors.

Wouldn't your customers love the following message? "We will always tell you what we do with your data. We make it easy for you to manage what we know about you."

Put yourself in the shoes of a customer – what may he/she be afraid of? If a customer really trusts you, that customer will not ask you to throw away what you know about him/her.

GDPR can also have some immediate hygiene effect internally: Analytics people often seem to translate the expression "garbage collection" from C++ or Java programming to data gathering: "Gather, store, and keep as much data as you can – it may be good for something in future."

Such behavior is ineffective. If you gather all data you can get hold of, you will probably have trouble determining the valuable part: You will miss the forest for the trees!

To use another analogy, you are creating *noise*: Think of your difficulties to understand single voices in a busy marketplace or railway station.

GDPR will give a tool at hand to enforce consciousness in gathering data. It will help an organization think and plan before gathering data unconditionally.

Finally, ethical behavior is not the enemy of success! If you make data privacy part of your organization's priorities, your long-term success consists of a superior reputation. Customers value credible behavior more than ever before.

# Recommendations

Here are my suggestions based on the preceding three aspects:

(i) **Address both systems *and* culture.**

Be clear that both aspects are equally important: Projects with clear deliverables AND a culture change.

(ii) **Extend data governance to data privacy.**

Properly governed data would make GDPR compliance easy – including the ability to demonstrate compliance.

(iii) **Add privacy to your organization values.**

Stress the opportunities that come with data privacy. Note that people will only follow this value if it is at least on the same level as other values. Otherwise, short-term benefits will be rated higher than long-term success factors such as how you treat data about your customers and other parties.

(iv) **Involve all stakeholders.**

As a CDO, you cannot do all of this on your own, and neither can a CIO. Cross-functional joint forces are required! Furthermore, all functional leaders in an organization need to understand both the risks of noncompliance and the opportunity of compliance.

# The Outside World

*"You know what you need?*
*A change of air!"*

**Figure 18-1.** Stew in your own data juices?

© Martin Treder 2020
M. Treder, *The Chief Data Officer Management Handbook*,
https://doi.org/10.1007/978-1-4842-6115-6_18

# Why look beyond my organization?

No organization exists in isolation. A lot of things that happen out there have an impact on the inner world of an organization. They form threats but also opportunities.

There are enough reasons to consciously follow what is going on beyond the borders of your organization. I am addressing some of the less apparent topics in this chapter.

# Sharing data across organizations

More and more organizations start sharing their data in groups, as part of what is meanwhile called "Shareconomy."

## Different ways of sharing data

You can share data in two ways:

(i) **Multilateral data sharing**

Several parties agree to codevelop a pool of data that all of them contribute to (usually by adding missing data, removing outdated data or cleansing wrong data), and all of them can make use of it.

To grant neutrality, third-party providers offer to do the orchestration of the data exchange. They offer methods to make data theft impossible. Examples of such providers are:

- "Corporate Data League" for customer and supplier data[1]
- "Skywise" for data used in Aviation[2]

(ii) **Bilateral data sharing**

This concept of sharing data is usually referred to as "data bartering". Two typical examples are:

- Two organizations agree to provide each other with data of more or less the same value, without charging for it.

---

[1] Founded by CDQ in 2016, collaboration with the University of St. Gallen; see www.cdq.ch/cdl-en

[2] Norman Baker, Head of Digital Solutions, Airbus: "Data management – what big data is doing for aviation," in *FAST Magazine* 2019.

- An organization offers a discount to customers who provide their own data in return.

Almost all types of data can be subject to sharing, ranging from address data to sensor data gathered in-flight.

# The motivation for sharing data

Why do organizations do this? Aren't they giving away valuable information for free?

In fact, you give, and you receive. And it is both about "more information" and "cleaner information."

The benefits are highest where competitors work together. They are obviously working with similar information, thus benefiting even more from each other's data.

But, if all of them win, where's the competitive advantage?

There are basically three aspects:

(i) **Competitiveness within an industry**

Selected organizations within an industry may share data to gain an advantage over other competitors.

(ii) **Competition between industries**

Even entire industries compete with each other! If, say, all railway operators improved their efficiency to the same extent, the entire rail industry would gain additional advantages compared to the trucking industry.

Or imagine all Express and parcel carriers collaborated by exchanging traffic data. The gained efficiencies (cost savings and faster deliveries) would make it more attractive for organizations to ship a spare part from location A to location B than producing it at location B. As a result, the entire transport industry wins.

One practical approach here could be a joint agreement in the area of 3D printing. The idea of sending standardized printing data over long distances and then printing the parts locally and trucking only the last mile is expected to become a more and more attractive alternative to intercontinental Express transport. It is cheaper, faster, and more environmentally friendly.

On this elevated level of competitiveness, the data-sharing organizations will, of course, continue to compete against each other. In short, they are jointly increasing the size of their market before competing for shares in that market.

(iii) **Competition for capital**

An entire industry can be made more profitable compared to other investments. It is a well-known phenomenon that organizations in the same sector impact each other's valuation – if one organization publishes bad results, other organizations often face falling share prices as well.

Now imagine the entire industry becomes more efficient. You can expect investors to consider all organizations in that industry more attractive.

# External data

## Internal data is not enough

What do the visitors on your website, your competitors' press releases, and Facebook have in common?

All of them provide valuable external data. Get it, understand it, merge it with your internal data, and you will better understand the world.

In order to be commercially successful, organizations need to understand all parties they are interacting with. A purely inward-looking perspective makes you blind to market changes.

Using internal data, your sales forecast would be an extrapolation of numbers from the past. Adding data from the markets will allow you to consider external factors that might have an impact on your future revenue development.

It is enlightening to realize that far more than 99 percent of what's going on is happening outside your organization!

This observation even applies to the big social media platforms. They, however, are very successful in asking their customers to keep them informed. Customers voluntarily share all their personal details with those platforms, teaching them how our entire world develops, as their customer base is big enough to represent a large part of the earth's population.

Most of the other organizations, however, do not have a sufficiently big footprint across all geographies and social groups to get a good understanding of "what's going on out there" based solely on information their customers are sharing with them.

# What you can learn from external data

Internal data is usually historical data, that is, looking back. Instead, you need data that helps you look forward.

Blending external and internal data helps you put your internal data into a broader context.

(i) **Learn about your competitors.**

Your competitors will never share their secret strategies with you (or with the public). Not even their shareholders may get to know.

But it would be good to know, right? Maybe they know something you don't know. And very often the key to success is not just to do the right thing, but to be the first to do so.

More easily than you would guess, you can derive the plans of your competitors from publicly available data:

- Assess their job ads to understand their future focus: What are they going to concentrate on, and where are they building up teams?

- Consider the history of their communications: While you can't tell their direction from a single message, you may be able to see a trend in a sequence of public messages throughout the year, where subtle changes in wording might indicate a shift in focus.

- Does a competitor shift investment from one area to another? You may not be able to tell from single investments – collecting all of them systematically may provide the full picture.

(ii) **Learn about your customers.**

Social media was invented so that your customers share data with you free of charge that they would never have given you on direct request.

That is at least what you could think after watching human behavior on the Web.

Especially failures on the part of your organization provoke honest feedback by your customers. Even if it wasn't meant to come as consultancy for free, that is what you can use it for.

Of course, these comments are not representative, as people tend to be more verbose about bad experiences than about cases that meet their expectations.

However, you can put them in proportion. How did the numbers change over time? How is your situation compared to that of your competitors?

Another advantage of external consumer data is that it is not limited to your existing customers.

You can learn about the preferences of all potential future customers:

- What do people need?
- How do they want to get it?
- How much are they willing to pay?
- What is their quality demand?

Changes in your observations help you discover trends early.

(iii) **Learn about your suppliers.**

What do other customers say about your suppliers? Most of your suppliers have a Facebook page that will tell you more about them than any rating agency. Following the right hashtags on Twitter provides a lot of insight as well.

Beyond that, you can, of course, learn about your suppliers the same way you learn about your competitors – the same recipes apply.

# The CDM and external data

## Challenges around external data structures

Should the Corporate Data Model (CDM) consider external data?

An organization may not be able to influence the structure of external data – but it needs to consider it in its CDM as far as it is relevant for the organization.

The main reason is that external parties (customers, suppliers, authorities) will not adhere to any organization's internal data structure.

As a consequence, external data needs to be mapped to internal information (i.e., the way your organization sees the world).

Examples:

- Your organization's internal split of the world into different countries may be different from the official set of ISO country codes. This is not straightforward! Even on the ISO level, the split is not unambiguous. Think, for example, of the Canary Islands: You can consider them part of Spain, ISO code ES, or you can label them IC (short for Islas Canarias), the dedicated ISO code for the Canary Islands. You have to expect some customers, suppliers, or authorities to use one code, and some of them to use the other code. Your systems need to be prepared, and the necessity for manual intervention should be reduced to a minimum.

- The quality of official postcode systems varies by country. While some of these systems may be perfect for the description of sales areas or the definition of service prices, postcode systems of other countries may be too coarse or have remained broadly unused. In such cases, organizations require their internal geographical split. These two ways of dividing the world into smaller geographic areas, however, should not be independent of each other. You may use the official postcodes, for instance, and break them down further by means of city or municipality names.

- Public holidays may differ from your organization's operational days. Imagine running a service facility in a country with many local holidays (e.g., Spain). This facility may serve customers in adjacent municipalities with different sets of public holidays, so that you may keep your facility open even during local holidays. This is the Reference Data you need to maintain so that systems know the operational days of each facility.

Third parties, such as your customers, suppliers, external partners, and even your employees (in the case of public holidays), will most probably refer to external data definitions.

In fact, you would not want to force those parties to consider your internal logic or your verbiage. That is why internal data structures should never be part of any customer offering.

Finally, customers order an end result, with defined quality and/or service levels. It is up to your organization as their supplier how to achieve it the best possible (and most economical) way. Customers should never be bothered with data your organization requires to get organized while being irrelevant for the customer to understand the product.

## Consequences for your data model

In each of the preceding examples, you need full flexibility to maintain your internal data independently of data out of your control. After all, you don't want to be forced to change your postcode-dependent sales districts just because the authorities have modified the tailoring of some postcode areas.

At the same time, the relationship between external and internal data must be well defined, so that your organization's software can translate between the two worlds. (And remember: In times of Reference Data, hard-coding is NOT a valid option!)

Furthermore, we need the flexibility to change how we operate without touching the product promise.

That is why you will usually need to map external and internal data structures comprehensively and unambiguously.

## The mapping of internal and external data

Unfortunately, there are two areas where mapping can become extremely challenging:

(i) **1:n mapping**

Where an organization needs to break down the external view further, for example, for internal processing reasons, you face a 1:n relationship. The challenge is that the externally provided data is lacking information to break it down automatically.

The three options you have here are the ability to determine the required information from other data provided by the customer (the data model should tell!), to ask the customer for additional information, and to determine the missing information internally.

You may need to enhance your CDM until all entities and attributes of the correct mapping are covered. And you require business rules which describe how to obtain missing information.

## (ii) n:m mapping

Sometimes you face mapping ambiguity in multiple aspects.

Remember the challenges around the logic of ISO country codes, as described earlier in this chapter.

And now imagine you want to map them to your *internal* country structure, where independent country organizations have their own P&L responsibilities for specific geographic areas. Now, think of the following two cases:

- You may have San Marino managed from your Italian head office so that it is part of Italy from your organization's perspective. At the same time, your customers there would insist on *not* being Italian.

- In politically disputed areas, revenue allocation should not depend on which country code a customer uses.

That is why complex relations between internal and external data structures require a precisely defined data model and mapping rules.

And please don't think this is a technical task. Content-wise, the business stakeholders need to drive the development. Let the respective Data Owner codrive the discussion. Data Architects are expected to ask the right questions and to translate business needs into the data model.

# Data Quality as a service?

Even if all of your data is validated and cleansed upon entry, it gets dirty over time. It is usually not the data that changes – it is the reality that changes, turning previously correct data into incorrect data.

Permanently keeping vast amounts of data clean is a tremendous task, and outsourcing this task may be a smart move.

But how do you determine what (and how far) to outsource?

Let's start with the objective of always having clean data at your disposal. I'd like you to consider three aspects.

Firstly, regardless of whether your organization has the capability of keeping data clean or not, it is usually not a wise decision for an organization to cleanse external data on its own.

Instead, any external data, unless accessed on demand, should be subscribed to so that it is kept up to date. Doing so is a matter of efficiency. It is obvious that one data supplier keeping data clean for multiple subscribers must be more efficient than numerous organizations cleansing the same data in parallel on their own.

Secondly, data controlled or administered by an external organization is better kept clean by that very organization. They have the authority, and they are familiar with their own data.

Thirdly, the best approach is to automate whatever can be automated, be it through algorithms or be it through external triggers.

Where possible, you'd go for fully automated corrections (such as address changes taken from external data providers).

In all other cases, your automated quality check (against Metadata and plausibility) would create alerts. As a second step, you would organize manual reviews and fixes of any reported records.

You can expect the number of offerings with AI-powered heuristics to increase. This will allow for machine preprocessing to mature further, leaving less and less manual reviews of suspicious records to your internal workforce.

Remember to exempt data with personal information from this approach (see Chapter 17's section "GDPR – All done?"). In all other cases of confidential data, AI-powered heuristics can definitely help avoid exposure of confidential data to external resources.

Wherever data cannot be fixed through a fully automated process, you would determine how and by whom the remaining manual work should be done.

- Where internal data is not confidential, and where its quality is well described, you could contract a process outsourcing provider to cleanse your internal data.

- In cases of data specific to your organization, the best know-how can probably be found in-house. But it is often not the obvious experts that may help you best, but the veteran caseworker who has been with the organization for decades.

- The same applies to data that requires knowledge of the local markets in which your organization operates. Unlike external providers, your local teams should have the necessary local knowledge.

It is a proven recipe to orchestrate the Data Quality work centrally while holding the local teams responsible. At the same time, you can demonstrate your trust in their knowledge and diligence.

A short formula for Data Quality improvement is

- Automate where you can.
- Outsource where there is a good offer.
- Take care internally in all other cases.
- Never miss out opportunities to improve Data Quality!

# Global standards

## Your own standards – good but not good enough

Data standards are good. That is why many organizations define internal standards to ease the exchange of information. As we have seen, those standards touch a broad range of Data Management areas such as data model, data glossary, and metadata.

A typical approach is to agree on standards, to map them to nonstandard implementations wherever possible, and introduce a repository where all mapping is documented.

The advantages are apparent. Your organization obtains the flexibility to migrate to those target standards in a phased approach. You may even include the integration of your customers' data, although you usually do not have any authority to drive changes on the customer side.

But in all of those cases, the impact is limited to your own organization and maybe a handful of cooperative customers.

In many cases, the success of a process across organizations will depend on *all* of your customers, suppliers, or partners to join in, that is, to speak the same data language. But why should they be willing to accept *your* standards? They may have additional effort mapping your standards to theirs. And they may run into issues with your competitors who are not likely to adopt your standards (unless your organization has the market power or authority to set industry-wide standards).

## Standards across organizations

So, how can you get closer to a seamless, mapping-free flow of data across organizational borders without having to convince others to adopt your standards?

### Go for global standards, wherever possible!

Your customers use standards as well. And particularly small customers may not have the bandwidth or the know-how to develop their own standards. They often adopt international, public standards from established standardization bodies such as UN/EDIFACT, ISO, or IEC. Your organization cannot be too big to follow this approach. Assess those standards, and you will discover that most if not all of them will work for your organization, as well.

You might find out that some of your customers or suppliers already use the standards you are considering. This reduces your migration effort even further, as it helps you simplify some of your external interfaces.

An excellent example of a global, company-independent standard is ISO/IEC 15459. This standard defines a globally unique identification mechanism, e.g. for serial numbers, package IDs, batch numbers and so on, which is a precondition for a seamless data flow along the entire supply chain, including traceability of the chain.

Unique identification allows for the physical supply chain and the corresponding data supply chain to be fully synchronized: Any physically exchanged unit[3] can be unambiguously referred to by a code wherever the data describing the transaction is transferred electronically.

This logic has been in use for decades, first standardizing the content of barcodes in the early 1990s. Today it covers the entire area of *Automatic Identification and Data Capture* (AIDC), including two-dimensional codes and RFID tags.

Its philosophy is similar to the identification of web addresses, where only the top-level domains require (very lean) central administration. Identification of parties on lower levels is maintained by the respective higher-level domains. ISO/IEC 15459 has it organized through so-called *Issuing Agencies*.

You can easily see the immediate benefits:

- Different Issuing Agencies can issue ranges of identifiers autonomously, without running the risk of any identifier being issued by more than one Issuing Agency.

- Organizations can exchange goods and identify them unambiguously without any need for prior alignment on formats or number ranges.

You find further information and a concrete application of ISO/IEC 15459 in two presentations I published on SlideShare several years ago:

- (Treder, License Plate – The ISO Standard For Transport Package Identifiers, 2012)

- (Treder, Basics of Label and Identifier, 2012)

---

[3]Examples are a transport unit, a product, an item identified by a serial number, or a reusable container.

# Beware of pseudo-standards!

Organizations, at times, develop "standards" unilaterally, and they hope that their mere market power helps establish those standards. These solutions are not sustainable, as competitors wouldn't want to adopt a standard they cannot influence. As a result, alternative, competing standards will emerge.

That is why you should ensure supported standards are maintained by bodies that fulfill the following three criteria:

(i) **Not-for-profit**

As soon as an organization earns money with the maintenance of identifiers or related services, it is no longer independent.

Unfortunately, nobody can prevent such organizations from calling themselves "nonprofit." That is why it is always advisable to have a look behind the scene. Sometimes the legal form of an organization reveals its commercial focus.

Another good indicator is the degree of marketing activities: If an organization is actively promoting its standards and services on various channels, you should be careful.

(ii) **Global**

In our globalized world, national standards no longer meet the requirements of unique identifiers.

Instead, you need to ensure that any identification standard works for cross-border business as well.

National Issuing Agencies are acceptable if they are members of an international standardization body and if they follow global rules to systematically prevent ambiguity.

(iii) **Independent**

Truly independent standards are maintained by independent agencies. Such agencies are always founded or tasked with the maintenance by supranational organizations.

Examples are the UN, ISO, or, in the case of the Global Legal Entity Identifier Foundation (GLEIF), the Group of Twenty (G20) and the Financial Stability Board (FSB).

In addition to these three preconditions, global standards also require proper governance, including rules for participation and escalation.

A mature governance model gives all users the possibility to coshape a system while not letting any single party dominate.

Such a setup allows for the community to ensure all necessary attributes of data are generally being taken care of, such as public availability or global consistency.

## Practical approach

But what can you do, practically, as an organization?

(i) **Understand existing standards and concepts.**

Entire organizations might not be aware of certain global standards. But even an organization's CDO might not be aware of standards already known at least to parts of the organization. That is why you should look around and talk to people.

I am a great supporter of a Data Office function that assumes responibility for external data. Such a team can also lead the stocktaking of available, independent standards.

(ii) **Promote independent standards.**

You can bring this topic up in discussions among your industry peers and with your suppliers and customers.

Wherever an entire industry agrees on an independent standard, they can avoid or minimize competitive disadvantages to single members. Instead, they can potentially strengthen their whole industry in competition with alternative industries.

For example, if the entire television industry made the usage of their services easier through standardisation, they could compete more easily with streaming services.

(iii) **Engage in the development of standards.**

How can you ensure global standards match your organization's operational needs?

Engage!

Global standards don't fall from the skies. They are developed and maintained by human beings. And as most of those standardization bodies don't employ full-time experts, they depend on people from all kinds of organizations to contribute.

If you decide that you or someone from your team should work on certain ISO standards, you should contact your country standardization body and ask for the right workgroups.

Those workgroups usually work toward continental workgroups. Finally, global workgroups meet and discuss, collect and review proposals, and come to final conclusions. It is extremely powerful and influential to be part of that process.

# Cloud strategy for data

## Outsourcing to the Cloud

Cloud solutions have become a popular option for data storage and handling at many organizations – for good reasons:

- Sharing infrastructure allows for flexible scaling, as demand averages out across multiple users of existing infrastructure.

- Noncore activities such as the operation of data centers can be run more efficiently by third parties who specialize on those topics.

- Data security, resilience, and compliance can be left to external specialists as well, leveraging economies of scale also in case of changes of regulation or external threats.

## Software as a Service

Cloud solutions usually come with SaaS offerings which promise similar advantages.

You might have observed that SaaS is growing fast where processes are relatively standardized across organizations, for example, Finance (SAP or Oracle) or Sales (Salesforce.com). Organizations easily outsource these tasks as they are usually not part of their core competencies or their unique selling proposition.

## Intelligence as a Service

The next logical step is: Where processes and logic get standardized, "intelligence" can be standardized as well!

Intelligence as a Service (IaaS) is not a new expression. It is probably as old as SaaS (Software as a Service) as the idea is basically the same.

But no globally agreed definition of IaaS has developed so far. One aspect most people agree to is that IaaS is a subset of SaaS.

In early references,[4] IaaS was often used to describe what we would more adequately call Reporting as a Service: Cloud services that created helpful statistics and reports based on provided data – in other words, something that all SaaS providers offer meanwhile on top of their functional software solutions.

But IaaS is growing beyond simple reporting. It is increasingly covering Artificial Intelligence (often also referred to as AIaaS) and Machine Learning.

This development stands for an aspect of SaaS that is gaining relevance. Very early, SaaS had started to claim your data: Now it is increasingly covering the business logic behind your data as well.

The idea seems to be: If creation, modification, and operational use of your data take place in the cloud anyway, why not doing the Analytics part "up there" as well?

Where external data gets standardized, the opportunity is growing further. You don't need to get all that external data yourself – an IaaS provider does it all for you, joining external and internal data in the cloud, finally providing you with desired insight.

This approach may even ease privacy concerns. An IaaS provider can aggregate personalized data before providing insight to organizations in an entirely anonymous way.

As an example, look at traffic data: The service Real-Time Traffic Information (RTTI) gathers data from individual cars. The resulting traffic information is entirely anonymous. Personal information is not necessary.

You will not hear me challenge any of the opportunities described so far.

However, the entire approach of Cloud, SaaS, and IaaS comes with risks that everybody should be aware of. Most executives are not.

Let me describe two of them: Data Model issues and black box issues.

---

[4]As an interesting example, see InformationAge (2006).

# Risk of Data Model issues

Organizations that work with more than one cloud provider (or with a hybrid cloud solution which you find even more frequently) may face this challenge: *Each cloud/SaaS solution comes with its own data model and data structures.*

SaaS solutions lack flexibility in their data model – by design. If each SaaS customer could have its own data model, a shared solution would simply not work.

Just adhering to all of those data models at the same time becomes logically impossible as soon as two different cloud solutions are not compatible with each other.

Here, the siloed character of functional cloud solutions collides with the cross-functional nature of any corporate data structure. Your SaaS solution for your Sales team has a Sales-focused view of what a customer is. This view most probably differs from the customer perspective of your Finance department's SaaS solution.

SaaS comes with a database structure that supports a single data model. There is usually some flexibility through configuration, but it is limited by design. Too much flexibility would make a SaaS offer impossible to maintain.

No organization can unilaterally enhance the data model of a SaaS provider's core solution, no matter how small (or how important) such a change may be.

And even where changes are possible, they will move you away from the optimized SaaS model, and you might run into issues with subsequent releases (which you cannot skip, by design).

The entire problem is not really new. Remember how much money got invested in adapting SAP solutions to organizations' specific needs, long before people started to talk about SaaS. The more you adjusted SAP, the more difficult (and expensive!) it became to maintain, the more functional surprises surfaced, and the higher the risk of missing backward compatibility with every new release.[5]

With SaaS, it becomes even more of a challenge. You pay the "carefree" model with loss of flexibility.

---

[5]For good reasons, SAP consultants have always recommended that you rather adjust your company's way of working to the SAP model than the other way around.

If you are willing to adjust your own data model to the SaaS provider's model, you will have to be prepared for the following challenges:

## (i) Backward compatibility

If you are forced to adjust your chart of accounts, how will you manage year-on-year comparisons?

And will you be able (or willing) to keep your old, difficult-to-maintain on-premise Finance system up and running as long as you need to keep pre-SaaS data available?

## (ii) Intra-SaaS compatibility

An organization can adjust to one cloud provider's data model – but how about *two* providers for, say, Sales and Finance solutions? What if they define "revenue" differently? What if their definitions of "address" differ dramatically?

Of course, you can run your functional processes independently of each other, and a lot of organizations do exactly that. But you risk a mismatch between the different data worlds.

---

## EXAMPLE 1

This real case has been observed in the transport industry:

Carriers need to consider higher transport costs for remote areas. This will need to become part of the pricing process for standard rate cards and key account discounts. Marketing may want to use this information as well.

To satisfy all of these needs, you need a sufficiently granular, unambiguous division of the world into non-overlapping geographical areas. And it must be identical between all functions! If an organization has full control over all of its systems, it can simply decide to do so.

But what if their IT solutions for the different business areas are managed by different SaaS providers, independently?

Some providers may work with postcodes only, others with a proprietary possibility to further divide postcode areas, and a third group may work with city and suburb names. Compatibility? No chance!

---

This can indeed be a huge problem. But you will also face several smaller issues in daily life:

- Your Finance SaaS may not handle the dunning process adequately as it fails to recognize that several unpaid invoices belong to the same customer.

- Rendered services remain unbilled as the Finance view of a third party differs from that of the Operations department.

- Segregation of duties (SoD) may fail as the respective solution does not recognize that the parties executing two complementary roles are identical.

Today, many of these cases remain unnoticed as it all happens somewhere in the cloud. Where issues surface, we tend to cure the symptoms as a root cause analysis is difficult (if not impossible) across all of these cloud silos.

It does not get better when it comes to Analytics. Which data structures are the Analytics folks going to use?

How does it work in a pre-IaaS environment? Whenever a Data Scientist has an issue or a question around the organization's data model today, he/she will walk up to a Data Architect and sort it out. This may even result in a change request against the current data model.

With IaaS, it's usually a one-way road: The IaaS supplier will inform you about the data model. That's it. No chance to review whether the model provides all details so that the data scientist can base all work on such a foundation. It is as it is.

Imagine you rely on your SaaS provider for Analytics as well. How will you ensure AI algorithms tailored for specific models will work with other models? Remember, you cannot change them as they are part of the SaaS offering.

Functional executives tend not to understand the risk of incompatible data models. They may not even know what a data model is, and this is fine in most cases. But even IT experts tend to rely on cloud solutions to professionally taking care of all your data issues.

How would you address the issue, then?

You need executive support, to begin with. That is why you first need to find real-life examples and share their business impact with the respective executives. If you don't obtain full executive support, all of your further activities may be futile.

Once you have secured your organization's support, you would need to address the governance topic: Any data managed in the cloud by one of your helpful service providers must be subject to your organization's data governance.

You need to understand the technical flexibility of cloud providers. Most of them offer less flexibility than they technically could. And you should force them to negotiate the solutions with a mixed team including Data Architecture, not just with the future business users and the technical IT experts.

Any cloud project must require the approval by Data Architecture. The Data Architecture team must be part of any such initiative from the very beginning, that is, during requirements gathering and vendor selection.

Here, they will need to validate that all business definitions of data entities and attributes – as well as the analytic models and definitions used by cloud-based business applications such as ERP and CRM – align with organizational standards. The Data Office needs to ensure that any identified gap is closed or documented as a concession.

But should an organization really put their destiny in the hands of a team of data specialists? Should it be possible for Data Architects to stop huge strategic initiatives? Not necessarily.

The critical point in this context is **transparency**!

Here is my high-level proposal for an adequate data governance setup – which you can even apply beyond Data Architecture:

- Any initiative, reaching from a small sprint up to a multiyear program, can only receive the green light if the data-related impact is clear and visible to everybody.

- Responsibility is with the Data Architecture team. Their disapproval can be escalated by the original requestor, to the Data Executive Board and finally the organization's Management Board.

- If the Management Board decides to accept a negative impact as they come to the conclusion that the advantages outweigh the costs of noncompliance, they may provide final approval.

- All of this needs to be documented as an Architecture Concession.

This governance setup leaves the pyramid of power intact, but it enforces conscious, well-informed Board decisions.

| DATA MANAGEMENT THEOREM #11 |
| --- |

**Effective Data Management enables conscious, well-informed decisions at all management levels.**

# Risk of black box issues

Notably, smaller organizations cannot afford to develop their own basis for AI. They have to rely on publicly provided Analytics interfaces, just as they have started to rely on SaaS.

However, several risks come with "information as a black box" and with the undisclosed logic behind it:

(i) **Improper usage**

You might overestimate the accuracy of the provided information. Or you might use the wrong service for the information you need. For example, you ask a route planning service for the correct address of an organization. That organization may already have gone bust, but the service provider doesn't know – because there is no reason for anybody from a bankrupt organization to inform address data providers about its closure.

(ii) **Dependency**

The market for IaaS may become an oligopoly. Offering services requires a considerable amount of data plus the infrastructure to keep it up to date. Not many organizations can do this.

Think of the usual suspects:

- Those who attract data science and can afford to pay them
- Those who are already providing the underlying SaaS solutions

(iii) **Wrong information**

Wrong data or logic, applied by one of your primary solution providers, will impact a significant percentage of users.

People will believe in it as they "have heard it somewhere else before" – although all of it refers to the same source!

It gets truly dangerous if governments misuse this situation to systematically misinform people.

## Options for proper IaaS usage

But what would be a good response? Think of the following options:

(i) **Consider your data model early.**

The world has seen too many cases where the decision for a particular Cloud solution was taken while the solution provider's CEO and the customer's CFO played golf.[6]

Of course, this was not a gut decision by the CFO. It was instead based on a lot of thorough evaluation work done by both organizations together in advance. Of course, the assumption that the Cloud provider would not have become so big and famous if its solutions weren't great may have helped get to an agreement.

But has an analysis against the data model been part of that assessment?

Remember, the data model is not an IT detail. It is a vital part of the formalization of your business model.

So, what is it you want to become part of the decision paper?

I am thinking of a gap analysis of the data structures, followed by a few technically possible options and the determination of the effort (plus risk) to move to any of these options.

These alternatives would range from adopting the vendor's model to adjusting the model to the current (or desired) organization model.

---

[6]Feel free to replace CFO by CIO, Chief Marketing Officer, etc. or to replace golf by dinner or the opera.

Going for the vendor's model in the SaaS solution while keeping the whole Analytics part in-house should be considered as well.

(ii) **Check your flexibility.**

You would have to discard your existing HR logic or your chart of accounts to go SaaS and IaaS. Okay, why not?

Remember, the SaaS solution (including its Analytics part) works well elsewhere. It may be different from your current logic but not necessarily worse.

As a rule of thumb, you can assume that it is acceptable to adopt a third-party model in areas that are **not at the core of your product offering**.

Your areas of key differentiation are probably full of unique knowledge about your organization. Of course, you want to modernize them technically (there are ways of migrating from Mainframe to SOA without giving up precious logic developed over the past 20 years). But you would hardly ever want to give up your competitive edge in favor of moving to a fancy Cloud solution.

But look at those supporting areas, those that simply need to work in order for your organization to be successful with its unique product offering: A Cloud solution will put these areas in the hands of people whose core expertise lies there. This move allows you to concentrate all energy on your core business.

(iii) **Ensure transparency.**

It has never been as crucial as today for the logic behind any publicly offered service to be known and visible.

This is particularly important for data web services – be it an operational query, be it a complex analysis.

In other words, if you don't know the logic behind a query, you should ask a lot of questions before using it commercially.

Thank goodness, the global Analytics community is a very open one, with a lot of willingness to share even the source code (mostly easily accessible on GitHub).

Public Domain is no longer bad. The danger behind it goes away with its transparency. Furthermore, the community is usually big and active enough to discover problems, fix issues and close gaps.

(iv) **Avoid vendor locking.**

No matter how much you like a certain provider (be it a SaaS provider, be it a data provider such as Dun&Bradstreet), you should design your architecture to stay technically independent.

You can expect, say, Oracle to complain vividly about your lack of willingness to use Oracle Analytics together with your Oracle Cloud solution – but there is no technical reason to give in.

(By the way, don't let those providers tell you that you face latency issues if you force them to work with data sources and data processing in other clouds. If long-distance data transmission were an issue, you'd already have faced that very issue when working with that SaaS provider and its globally distributed data centers in the first place.)

Summary: You better ensure you stay in control. The Data Model is yours, not that of Oracle, Salesforce.com, Workday, and the like.

And Analytics is yours – the solution providers may offer you the technical platforms and fancy visualization tools, but you should own the logic and content.

# Blockchain

A lot has been written about Blockchain and the opportunities that come with it. I don't want to repeat any of this. Instead, I'd like to share a few thoughts you may face during the implementation planning.

## Unique identification and blockchain

Blockchain seems to be becoming the primary technology for decentralized, change-protected logging of transactions and other information that relates to more than one party. Contractual partners of any kind can just add their contractual information to a blockchain and have it signed by thousands of other, independent parties. Those parties cannot confirm the correctness content-wise, but their signatures allow for the determination of whether the record has changed or not.

Great stuff!

But how do the involved parties secure the meaning of identifiers used in that record? If party A later claims that "product 12345" was something totally different from what party B assumed it to be?

An approach can be to store all of that information in the blockchain as well. But this is, of course, highly redundant information. If an organization offers a product with a description of several megabytes, and it sells the product 1 million times, the blockchain is massively inflated.

One solution is the creation of a separate product blockchain where organizations link product documentation (including all changes over time) to unique identifiers so that no unauthorized change can happen without evidence.

But how do you identify items unambiguously without again requiring a centrally managed data repository? A number 123456 could be a serial number, shipping number, order number, and so on of organization A, B, and C. To become unambiguous, you'd have to add the organization identifier. But where do you take that organization identifier from? From another centrally maintained repository? This may again not be a universally comprehensive one and thus not contain all organizations on earth. It may be a commercially managed one and therefore dependent on an organization that has its own interests and could even go bankrupt.

So, what's out there that can be considered at least sufficiently neutral and global at the same time?

My response to this question is the recommendation to work with global standards, as described earlier in this chapter.

In fact, ISO's concept of globally unique identification of goods, of packages, of reusable containers, and a few other entities can help us here as well. Distributed logic without a central authority or a prespecified number of participants benefit from standards that don't require any formal agreement between all players.

# A false sense of security

A Blockchain ensures that any retrospective change to a transaction becomes visible. In reverse, if the Blockchain says it's all okay, then you are safe.

Right?

Well, technically...

Honestly, this is true if you understand the algorithms and if you know the source code.

But put yourself in the shoes of a user – be it an individual using it for transactions with unknown partners (such as trading Bitcoins), be it an organization's controller who uses Blockchain to secure business transactions.

That user will be using some sort of software to add entries to the Blockchain – it is highly probable that the user's normal working software does this in the background, as an additional step. And this step may even be hidden from the user, other than a message popping up telling the user that "the transaction has been secured and cannot be changed anymore."

This software, however, may be a malicious software, or, more probably, it may be a legitimate software that was hacked by someone. Do you think the end user could tell the difference?

And "hacked" doesn't necessarily mean the software is now permanently doing bad things. Instead, it may work exactly as expected, except for the small number of transactions you do with one particular business partner. It may create syntactically correct entries with incorrect content to the Blockchain, and everybody would be able to see the deviation, unless… yes, unless the malicious software goes a step further and tells the user that everything is okay. How could that user tell otherwise?

Have you ever checked the content of a Blockchain yourself, other than through a piece of software? Do you trust in that piece of software? Where does it come from?

Okay, you say it's about certifying the software? Yes, that is an important part. But who is certifying? Remember, there is no central, neutral authority. Do you know the issuer of the certificate sufficiently well to trust that body?

Furthermore, if your software tells you it's certified, how do you know it's true? Can you trust in your software?

These questions demonstrate that an organization adopting Blockchain will need to do more than just defining the business logic and implementing the logic in the software (or maybe even only using software that has it all built-in already).

But what would you do on top?

First of all, the technical knowledge around Blockchain is nothing an organization should outsource in its entirety. You need experts who really understand what is going on, rather than relying blindly on "black box" Blockchain services.

Secondly, these experts need to be able to look at the source code, ideally together with information security specialists.

Thirdly, different experts need to look at the same solutions independently. Otherwise, the one single Blockchain guru in your organization may turn out to be part of the gang.

Finally, software should monitor software. If you use off-the-shelf software, SaaS, or public domain web services, you may wish to add your own routines to check for suspicious results.

Overkill?

Remember it is becoming more and more challenging to discover manipulations in this area through common sense. This makes this area extremely interesting for criminals. And they may not clumsily have the whole system collapse one day – they will rather try to keep their intrusion undetected.

The situation is a bit similar to that of the first-generation banking software where single programmers were able to add logic to transmit micro-amounts to their own bank accounts. They were able to sustain this for years, making the programmers rich without anybody missing any money.

It is, therefore, a good idea to anticipate such behavior from the very beginning when entering a new era with new technology.

# Handling Data

*"No, I'm afraid we can't 'just make the data up'*
*—this is business, not politics..."*

**Figure 19-1.** Data handling in business is serious

© Martin Treder 2020
M. Treder, *The Chief Data Officer Management Handbook*,
https://doi.org/10.1007/978-1-4842-6115-6_19

# The Virtual Single Source of Truth

## What does a VSSoT look like?

Do you consider your organization's data lake your Single Source of Truth? Or at least the target?

You shouldn't.

"Single Source of Truth" should **not** be considered the bucket full of golden data at the end of the rainbow.

Instead, data is always on a journey, from creation or entry into the organization throughout modification up to consumption and dismissal.

Data may be collected, structured, transformed, merged, or joined. Each new instance, be it a table, a database, or a view, should become part of your Virtual Single Source of Truth (VSSoT).

Remember, creating such a new instance does not render any of its predecessors invalid. The same data can be found in different tables and views, and picking the most adequate source is not always trivial.

Unlike other assets, data does not reduce through consumption, so that it can repeatedly be modified and used in different forms and composition, without diminishing. Modification is not even a linear process. Think of operations like Outer Joins where two tables together create one new table.

Furthermore, different evolution stages of the same data can coexist. This may be helpful, but it comes with a risk: Consumers may accidentally use the wrong evolution stage, for example, a version of the data where certain records have been filtered out or aggregated, for the result to serve a different purpose in another area.

A typical case: "We have always used table X from department A for our analysis. Nobody told us that department A changed their logic to create table X one day."

That is why **the entire data lineage** needs to be covered when talking about a single source of truth:

- Creation or acquisition of data needs to be defined unambiguously (origin, rules, timestamp).

- The journey of data from one repository or stage to another must be well defined. This includes the timing aspect, to avoid inconsistencies through different speeds of modifying different data elements.

- Any manipulation of data needs to be done in a well-defined way, as long as the output is made available to multiple consumers.

- Aging of data needs to be well defined as well: As of when is data considered "outdated"?

You can imagine the complexity of your VSSoT if you expand Figure 19-2 to match your organization's data lineage. To which degree is all of this being documented within your organization?

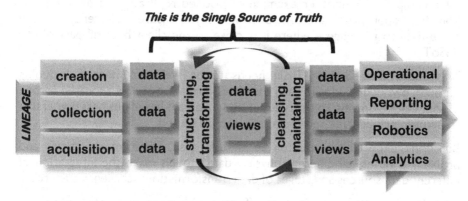

**Figure 19-2.** The complex Single Source of Truth

## How do you shape a VSSoT?

Working toward a Virtual Single Source of Truth starts with stocktaking.

- Map the flow.

- Add the time dimension (sequence; real time vs. lag; individual vs. batch).

- Understand the purpose of each data transformation.

The result will be a "virtual data source." In an initial phase, you can transform it into your "Virtual Single Source of Truth," before you will start optimizing it in a second phase.

Such optimization consists of the assessment where data should be centralized, but it is also about synchronization of data flows. However, it is **not** about aiming at building the *one single database!*

Even replication of data can be compatible with the single source of truth principle – if all "copies" are managed as part of one single source (and changed synchronously).

The physical structure of a data repository, for instance, needs to support the consumption. If the same data is to be used for entirely different purposes, it may need to be replicated to physically different repositories, for example, in-memory databases or highly denormalized table structures for real-time operational purposes versus Hadoop clusters or Spark Resilient Distributed Datasets (RDD) for massive parallel analytics processes.

Edge computing is another example of justified replication of data – to be done in an appropriate and controlled way: If you ensure consistency, you can move data to the edge – where it is close to the client but still part of the VSSoT!

This entire Data Supply Chain needs to be well documented, including business ownership. Ideally, each data transformation step is allocated to one of your Data Domains, so that your existing, domain-based business ownership structure, including related processes, can be reused.

But don't expect all of this to be an easy task. Different structures and probably a variety of different physical data models make the integration of different data sources and stages of data transformation a challenging exercise.

And please don't consider it a one-off activity. New requirements lead to new transformations, and this vivid ecosystem requires active management.

It is for good reasons that, during a conference in 2018, Rick Greenwald, Research Director of Data Management Strategies at Gartner Group, predicted that "data integration will become the most time-consuming task in future."

# Single source of logic

## Service-Oriented Architecture (SOA)

Okay, you have all of your data under control. While it is maintained locally, there is one central (logical) place for your data, creating a Virtual Single Source of Truth…

…for your data.

But how about your logic?

Do you leave the implementation of logic to the client applications? Do you allow them to take perfect, unambiguous data from a single source of truth just to let them implement the logic independently of each other?

Such an approach can easily lead to the same issues as with data that is maintained in different places independently.

My preferred solution is based on the good old Service-Oriented Architecture (SOA).

You might have learned a couple of advantages that come with the SOA concept, including the possibility to work with small, independent development teams through clearly defined request/response interfaces or the support of an agile approach that comes with that independency. The single source of logic is another, often underestimated advantage of this approach: Not only does everybody use the same raw data – no, any operation applied to data can be standardized as well.

As you need fewer people who have to implement the logic in their respective applications, you reduce the risk of different interpretation and implementations. Testing becomes easier as you don't need to test multiple applications in detail.

## Distributed logic – the Octopus Principle

SOA usually requires direct real-time access to centrally provided web services, as illustrated in Figure 19-3.

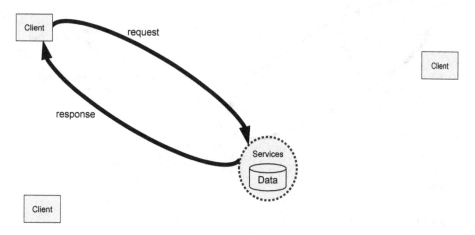

**Figure 19-3.** Web service principle

But what do you do if this is not possible?

Think of those customers or business partners you expect to use your web services:

- They may not be willing to accept the risk of data connectivity issues disrupting their supply chains, as parts of their software application lies on the other side of the disruption.

- They may be afraid of relying on software that is out of their own control, run on another organization's server. (After all, you are not Salesforce or SAP but an "ordinary" business partner.)

If an external partner reacts this way, would you return to the old way of working? Would you again send them your data, including regular updates, and ask them to mimic the logic you implanted internally? In other words, back to Stone Age ways of working?

The solution to this is NOT to send raw data files to the client application again!

Instead, you can have the web service move close to the client, as you can see on the right side of Figure 19-4.

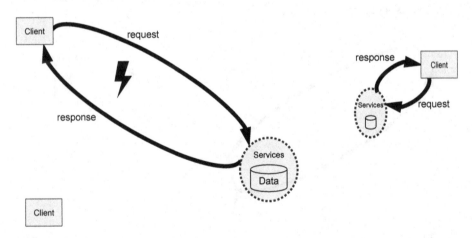

**Figure 19-4.** Local web service

In practice, you would provide your business partner or customer with a piece of software that acts as a black box, doing precisely what your internal software does. Ideally, it would be based on the same source code.

As a second step, you'd ensure that updates of both data and software are controlled by the service and that they are done in the background.

As a result, the entire parent service is mirrored to the child services, as shown in Figure 19-5.

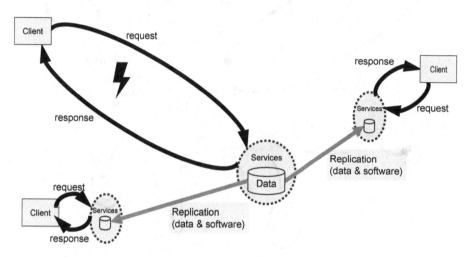

**Figure 19-5.** Data replication to local web services

This approach mitigates the risk of connectivity issues or failures of your own infrastructure. The worst thing that could happen is nonavailability of the update controller, resulting in the local web services operating on slightly outdated data.

Now imagine this mechanism being applied to multiple local servers for web services, run at various business partner sites: This is the **Octopus**, with its tentacles symbolizing the ability to bring data *and* logic close to the client (Figure 19-6).

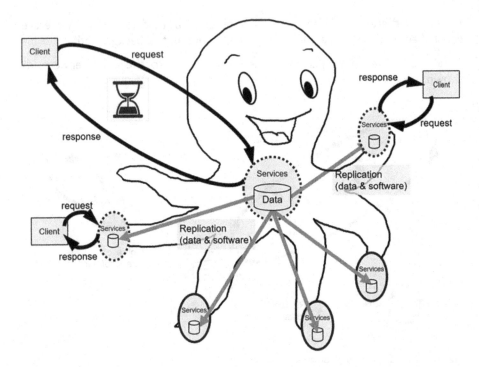

**Figure 19-6.** The SOA Octopus

This is the most critical aspect of the "Octopus Principle": The octopus represents your Virtual Single Source of Truth (VSSoT), with its tentacles reaching into edge solutions. It is no longer the edge application that is responsible for keeping its data and logic up to date. In fact, the entire virtual SSoT can (and should) be totally opaque to those applications.

The edge application doesn't even need to care about whether it calls a local or a remote instance of a web service. The interface is identical, and the differences are limited to IP address and port number.

Edge repositories may be restricted to subsets of the central SSoT, both in terms of attributes and records. A product ordering service running in a particular country, for instance, does not need to replicate the price lists for all other countries.

But the octopus itself needs to ensure all data selected to be replicated to the edge repository is kept up to date and that referential integrity is ensured.

Similarly, the registry of the SSoT, that is, the entire map, needs to be kept up to date. Its creation is not a one-off activity. It needs to become a permanently up-to-date source of data for each stakeholder in search of the right information.

The time dimension plays an important role, as well: All data needs to be available in time for any client. Delays would often lead to inconsistencies between two interdependent data sources.

However, data does not need to be there earlier, either. In some cases, you might be able to save money by deferring data flows.

To find the most adequate provision of data to a client site, it is helpful to incorporate the risk of wrong or late data into your business case. Sometimes it can become seriously expensive, for example, if a customer orders against an outdated price list or if a customer's supply chain is interrupted due to crucial data that has not yet been replicated to the customer's site. In other cases, it does not really matter, for example, where data is used to roughly indicate a performance level and where a small fraction of outdated information does not impact the result substantially.

Where the overall impact of slightly outdated information is limited, you could even plan for a regular offline operation, reducing dependencies on data connectivity and updates even further. The same applies to Reference Data that does not change frequently.

Finally, a sophisticated update mechanism would update the *logic* to any local instance as well. Just as with most third-party software on your PC, a local instance would regularly check whether it is still up to date. Whenever a newer version is available, the local client can download its successor and restart with the updated logic.

Everybody talks about edge computing today. You can consider the Octopus Principle an enabler of edge computing, with replicable results.

# Configuration vs. standardization

A variety of different meanings of the same expression is obviously not a Configuration vs. standardization good idea. Multiple logical data models covering the diverging needs of various business units or geographical units of an organization create the same issues.

But is standardization always the answer?

In several cases, there are excellent reasons for different logic or different definitions.

This is where I recommend going for **configuration**: You allow for different meaning or different logic under the same data model.

Let's have a look at two examples covering frequently used data domains.

---

## EXAMPLE 1

**Products:**

Think of an organization that operates globally. Based on regulations, it requires deviating product specifications in some of its markets.

Instead of considering them exceptions, those deviations should be modeled as variants of the same product, depending on the respective markets.

---

## EXAMPLE 2

**Facilities:**

Any big corporation has to recognize and identify different types of facilities, be it office facilities, warehouses, or construction sites.

However, none of these different types of facilities should be looked at in isolation, as many attributes are relevant to all facilities.

That is why there is a massive opportunity in having all facilities identified following one single logic.

You would, however, add attributes such as facility type, so that different applications can have their specific views on the same set of data.

---

Doesn't all of this make the data model more complex? Yes, it does! But remember, the data model is supposed to reflect the business model. If such a data model is too complicated, your business model may be too complicated as well.

So, yes, why not ask for standardization? But please start with the business model, not with the data model. If the data model indicates the need to act, you can assume that your business model is too complicated, and you should start there.

If the organization finds the level of business complexity adequate, however, you cannot expect it to be reflected by a simpler data model.

The data model might be used to shape a harmonization road map, considering both current business needs and technical backward-compatibility requirements. Even if the business logic is fully updated across all systems, you might be stuck with complexity for historical data. And you may face challenges with comparing data year on year. But this is the inevitable price of change.

# "Effective Date" concept

An "Effective Date" replaces a single attribute value by a historical stream of values for the attribute. Each of these values has an "Effective Date" and an "Expiry Date." Some additional rules:

- Whenever a value is to be changed, the old value is expired with the last day (or hour, or minute, etc.) of validity, and a new value is created with an "Effective Date" directly after it (i.e. with no possibility for a time stamp to fit inbetween Expiry Date and Effective Date).

- If a value is no longer valid, it is not to be deleted. Instead, it is kept, and an "Expiry Date" is allocated to it, indicating its last day of validity.

- If no expiration is foreseen, the youngest value of the chain needs to have the "Expiry Date" set to a value that represents "eternal."

The need for consistency requires all changes to be applied in full synchronization. Without an "Effective Date" concept, a postcode area may be allocated to a sales territory that does not yet (or not anymore) exist. Maintenance starts with the independent data fields, followed by fields depending on them.

But why do we need the Effective Date concept? Not all data changes can be applied within a second at midnight. An "Effective Date" concept solves this issue by allowing for modifications to be configured in advance.

If combined with a properly implemented Service-Oriented Architecture, it makes it possible to apply consistent changes at any time. All new values become active during the same second, and consistency is guaranteed.

Maintenance can be done through web services that hand over the target value, "Effective Date," and "Expiry Date." The web service itself does the technical work, that is, it does the necessary validation checks, and then it inserts the new value into the existing chain. This would usually include the change of "Effective Date" and/or "Expiry Date" of current attribute values.

This approach even allows for the handling of ranges (e.g., a range of postcodes or numbers): If a part of such a range is supposed to change, we can leave it to the web service to figure that out. It may split the entire range into the part below the subrange that is subject to change, that subrange itself, and the range above. While the ranges above and below inherit the original chain sequence, the center chain would get the new value inserted as described earlier.

Usage can be done in two ways: Either the client asks for a snapshot of a given moment (the default moment would usually be "now"), or the chain is requested, optionally providing a start date and an end date (if none of both is provided, the entire chain is returned).

This also allows to reestablish the status of any time in the past, across all data sources and data repositories.

In Analytics, the "Effective Date" concept allows for a Masterdata snapshot at any given time.

# Making data international

## The Babylon effect

Unless you work for a strictly national organization, you will have to deal with multiple languages. This challenge applies to suppliers, authorities, and customers.

Customers, in particular, are not willing to accept an organization-defined standard. They will always insist on their own language to be their communication language when talking to your organization. In most cases, there will be competitors for which this is not a problem.

Of course, you could simply store all versions and process them as they come. However, your warehouse staff in Alabama or your production folks in Bangladesh may not understand a local expression of a country somewhere in Africa.

There is no easy solution. Always translating everything from the source language to all other languages comes with enormous effort. Furthermore, you will quickly run into ambiguity issues, as there are multiple valid ways of translating from language A into language B.

## Alias terms

After you have translated a local expression to your internal version and stored it in your Masterdata, you'd want to be able to compare it with the same expression whenever it is entered by a customer online. This can only work if the entered information is translated precisely as the original term was translated.

A first solution approach is to define a leading language and to store all "primary terms" in that language. All local versions would be added as "alias terms."

Any version used by a customer can then be identified and translated to the "primary term." This version is to be used across your organization, by all employees who don't know the local expression.

As an example, you may wish to store all city names globally using their official English translations. You would translate Wien to Vienna, which becomes your "primary name." Whenever an Austrian customer enters "Wien," a lookup returns "Vienna" as your internal "primary name."

Other examples are product names, legal expressions, or country names.

# Transliteration

But how are you going to deal with different alphabets?

While the world's main six alphabets (Latin, Chinese, Devanagari, Arabic, Cyrillic, Dravidian) cover 90 percent of the world's population, you need to support the top 20 alphabets if your organization has a truly global footprint. And most of these alphabets have various variants that need to be considered.

This is where transliteration enters the scene. It defines the transition from one alphabet to another, character by character. Where the target alphabet is the Latin one, the approach is also called *Romanization*.

Unless you extend the transliteration by adding marks to the target characters (diacritic characters), the way back to the original word may be ambiguous – but this may even be inherent to the source language's spelling, which means string comparisons (as used in IT) are not safe anyway. As a consequence, you'd always store the transliterated version, and before you compare it with a provided word, the latter is transliterated as well.

This approach obviously requires a standardized way of transliterating. You would immediately think of ISO standards, and this is the right way to go.

However, even ISO standards are ambiguous at times, and sometimes you'd need to know which language a provided word is meant to represent, in order to select the adequate ISO standard.

That is why each organization needs to assemble its internal standard, to obtain reproducible results.

A challenge for the automation of transliteration is its difficulty with languages not based on single characters. Most prominent examples are logographic alphabets such as Chinese (where each symbol represents a word) and syllabary alphabets such as the Japanese Kana (where each symbol represents a syllable).

In some alphabets, the transliteration of one character depends on the following character(s) – which makes it nearly impossible to transliterate on the fly, for example, while someone is typing text into a data entry screen.

Consider that even ISO offers different standards for the Romanization of Russian Cyrillic or Japanese text. In each case, you need one internal standard for doing it, which allows for each word to be stored in exactly one Latin version. Each data entry of that word can then be romanized immediately and compared to the Latin reference version for a possible match.

All in all, the topic is complex. But if your organization works in many different countries, including a cross-border aspect, you will have to address it. Thank goodness, there are solution providers to assist you in doing so.

## Country-specific languages

In your process of getting ready for the interaction with customers from various countries, please resist the temptation to oversimplify, by just selecting the assumed closest match among the languages you are preparing.

People often prefer speaking English over speaking a language close to their own – which often stems from the disliked neighbor country. Here are two typical examples:

- Ukrainian usually understand Russian, but they do not appreciate being forced to communicate in Russian (and, by the way, Ukrainian Cyrillic differs from Russian Cyrillic!).

- French differs between France, Belgium, Canada, and the French part of Switzerland. Some of these countries are stricter than others when it comes to avoiding Anglicisms.

That is why, at a minimum, your international setup should always be based on the combination of language and country.

Luckily, many organizations have been facing similar challenges so that a set of de facto rules and standards have emerged, and respective public web services are available.

Whichever source you decide to select, you should make it an official rule so that your organization's Internet (and Intranet) presence, processing algorithms, and storage are all working consistently.

# Data Debt Management

It is an illusion to expect an environment where all projects are always aiming at getting it "right first time." And, indeed, in many cases, there are good reasons to start with a workaround, for example, if you need to quickly close a security leak.

Do you have to give up on compliance with essential data standards in such a case?

If you do it well, you don't – at least not in the long run.

The idea behind Data Debt Management is to only compromise on the time dimension.

In practical terms, you would insist on a data approval to be required in ALL cases, including those "urgent" ones. In reverse, you would specify that a "temporary approval" will be granted if the proposal itself already contains the commitment (plus funding and resources!) to fix the noncompliance by a specified deadline, usually in a second phase.

There may be little reason to apply this universal approach to data-related compliance only. It should be a general approach to avoid long-term workarounds. (See also under "Technical debt handling" in Chapter 8)

Organizations following an agile approach will be familiar with the concept. However, experience has shown that especially those organizations are least strict in following up on architectural debt and in enforcing the closure of gaps as per the agreed timeline.

Tolerance in this area will inevitably lead to a fast-growing debt pile!

Here is a possible recipe to formalize your Debt Management process:

(i) **Set up an Architecture Review Board.**

Have an Architecture Review Board with clearly defined membership, covering all Architecture disciplines:

- Technical Architecture

- Data Architecture

- Business Architecture

- Solution Architecture

Note that while responsibilities for Solution Architecture may be shared across different teams, this body needs to have the final say.

(ii) **Introduce a mandatory review process.**

Proper Debt Management requires an Architecture Review Process.[1]

You do not only require such a process for business transformation initiatives. It needs to cover everything from your organization's big change initiatives to business as usual, and it should take into consideration that changes in the architecture landscape may render previously compliant solutions invalid.

To begin with, the process should describe all trigger events (i.e., those events that initiate the process). The most frequent trigger event will certainly be the submission of a project request, but requests for clarification or change requests to approved projects should be covered as well.

The precise list of necessary information should be included, ideally in the form of a checklist. Remember to keep the list as lean as possible to avoid any bureaucratic overhead. Mandatory fields are the suggested deadline for closure and the department in charge.

The main task of the review process is to describe the handling of discovered architectural gaps. First of all, the process needs to contain criteria for the acceptance of temporary workarounds, for example, current violation of public regulations or significant revenue losses.

As with every process, responsibilities need to be clear. Who can approve, disapprove, or override a previous decision? What are their (dis-)approval criteria? What is the escalation path?

(iii) **Add an "Architecture Debt" process.**

After an exemption has been approved by the Enterprise Review Board, a systematic follow-up is indispensable. You should cast it in a well-described process.

---

[1]If such a process is already in place, it should be reviewed for the requirements described here.

To begin with, the process needs to ensure all (approved or detected) deviations from Architecture standards get documented. It should be based on an unambiguous identification logic, assigning a unique identifier to every single exemption. And the approved "Expiry Date" must be documented for each exemption.

The main part of this process is the description of a proactive follow-up process for all exceptions, from approval to closure.

Finally, this process should specify the responsibilities for all necessary activities. The Data Office needs to play the primary role in all data architecture matters.

(iv) **Ensure full transparency.**

Architecture exemptions should not be dealt with behind closed doors. Everybody should know the compliance status of every initiative, process, or solution.

This is not about bothering the workforce with technical details. Instead, you would think of well-defined metrics so that both the level of architecture compliance and the degree of adherence to the approved deadlines could be read from regularly published KPIs.

The necessary visibility should be achieved through ongoing communication. Furthermore, general training should allow all relevant people to know the process sufficiently well to be able to follow it.

# Agile and data

## A solid foundation – why?

Should you go Agile or Waterfall when dealing with data?

I fear this may be the wrong question...

There are good reasons to avoid an either-or decision here.

Whenever you need to decide about the project approach, you might wish to ask: What can easily be changed in an agile way, and where is it better to do careful preparation and diligent planning?

At first glance, Agile is the way to go, as it avoids a lot of issues that had come with the waterfall mentality, particularly the setup of siloed responsibilities: You implement as specified even if you saw that the specs didn't make sense. Nobody could blame you for implementing according to those specs, but if you deviated for the best reasons, your head was on the block if anything went wrong.

Furthermore, Agile encourages an organization to change direction whenever necessary. This is critical to avoid what London Business School calls "Escalation of commitment": You keep doing something, although you know that it is wrong.

But given the need to respond quickly to an ever-changing environment, is time-consuming prework really so important? Or is it wiser to always shoot from the hip? Should you take Data Architecture decisions in an agile way that you might be logically tied to for the foreseeable future?

Einstein once said: "If I had one hour to save the world, I would spend fifty-five minutes defining the problem." He didn't say 15 or 30…

Einstein had good reasons to put it this way, and they are still valid today.

Here are three strong reasons for a structured foundation based on planning:

a) **It makes you agile in the long run.**

b) **Ad hoc decisions create interdependencies.**

c) **Quick fixes risk curing the symptoms instead of addressing the root cause.**

If you prepare for the unexpected, you will be able to react more quickly when it comes. Two examples:

- Even the most agile approach will not allow you to change the database schema across multiple applications and interfaces quickly. And agile changes of message formats are particularly complicated where EDI with suppliers and customers are involved.

- Missing data cannot be recovered if recovery has not been part of the original considerations.

Does this mean that Agile and Planning contradict each other?

No. Agility and sustainability are two important targets you will have to keep balanced.

A good example is the agile concept of technical debt that allows you to proceed quickly while ensuring that noncompliance does not get forgotten.

As part of an adequate data process (see section "Project data review process" in Chapter 8), it provides you with a tool to balance the need for quick implementation against a long-term perspective.

## A solid foundation – how?

To become Agile in the long run, you would want to put the following components in place (most of which have been discussed in this book):

- Web services that applications can connect to within minutes, ideally through an actively managed enterprise service bus.

- Documented knowledge of where the information is.

- A structured change process (knowing whom to ask).

- A corporate glossary, removing the need for initiative-specific glossaries.

- Clear data guidelines, to be communicated before any activity starts.

- A Corporate Data Model which makes it easy for IT to implement – they can derive their database schemas directly from the business data logic.

- A self-sustained Six Sigma culture where people voluntarily select Six Sigma tools like Gemba Walks, 5 Whys, or Fishbone diagrams to discover the root cause (Agile does not mean "curing the symptoms"!).

All of these aspects need to be part of your data governance as most people in an organization would intuitively prefer quick gains over such a sustainable approach.

# Starting with the happy flow?

## What makes an initiative successful?

Whether you work in an Agile or Waterfall mode, you want to finish your initiative quickly. And there is nothing wrong with this aspiration. After all, the earlier you finish, the sooner the benefits kick in.

But which price are you willing to pay?

A common approach is to go for the MVP, the "Minimum Viable Product." However, besides the fact that such an MVP is considered the smallest scope that makes sense in production, people don't share a standard view of what it entails.

Unfortunately, a common interpretation is that of the MVP as a solution that covers the "happy flow." In other words, it works well if all rules are adhered to.

The advocates of this definition suggest that we should first concentrate on the happy flow. Once it is in place, we can look at how to deal with exceptions.

The background of such statements is usually not a profound theory of maximizing efficiency or a thought-through way of investing money, but simply time pressure in a project.

Based on my experience with hundreds of processes, I have come to suggest the **80/5 rule**: If you want to cover 100 percent of all cases, 80 percent of the investment is required for the ability to manage the 5 percent deviations from the happy flow. (It will never be exactly 80 percent or 5 percent, but the order of magnitude applies to most cases.)

A second observation says that correcting those deviations causes the lion's share of business costs – either directly through the higher effort to finally reach the target outcome in those cases or indirectly through impacted customer satisfaction and ruined reputation.

That is why the equation "MVP = happy flow" does not have a good business case.

But do you need 100 percent coverage? Probably not. I mentioned the famous 80/20 rule earlier: Covering the most beneficial 80 percent of all functionality with 20 percent of the effort is more economical than covering the full set of business requirements by spending five times the money.

Similarly, when looking at possible deviations from the happy flow, you may be able to cover 4.99 out of those 5 percentage points for half the cost of covering 4.999 percentage points. This is a lot of money for such a small increment!

Again, don't focus on the exact figures. Instead, find out whether the *business impact* of an issue justifies the investment to prevent it from happening.

## The three dimensions of success

Eventually, if you want to reach the most adequate solution, you'd need to base your scoping process on facts and data, instead of individual feeling ("My experience tells me...") or a misdirecting rewarding scheme ("...delivering on time is the main thing...").

We talked about conflicting targets within the organization in Chapter 16. Project managers and product owners face a similar dilemma.

Any initiative has to optimize between **costs** ("Stay within budget"), **speed** ("Deliver within schedule"), and **quality** ("Meet all requirements").

How will a project manager balance these contradictory requirements? After all, the academic approach of creating an adequate weighted target function is neither straightforward nor does it provide the project manager with an implementable plan.

Instead, the project manager will prepare a proposal for the business owners, hoping that they will consider it sufficiently balanced. This is a fair approach.

But what happens if something goes wrong during the initiative? Even an agile approach may run into severe issues, for example, if a selected technical solution turns out to be insufficient or if it takes very long to find a software bug.

Such cases require a revision of the three dimensions: Do we increase the budget, do we allow for more time, or do we accept a reduction of quality (i.e. compromising on business requirements)? In any case, the project manager will feel unhappy – nobody likes to present such a story to a steering board, even if there are excellent excuses.

So, how could a project manager avoid having to admit failure?

- Budget? In mature organizations, it is difficult to informally increase the amount of money to spend. Some spare buckets may have been prepared as a matter of prudence, but options are generally limited.

- Time? As soon as a schedule has been agreed and published, you cannot change it unilaterally anymore. This applies to both Agile and Waterfall, even if the former allows for deadlines to be moved as part of the methodology.

- Scope? Features that have been agreed cannot be taken off the list (or moved to the product backlog) without business consent. The same applies to nonfunctional requirements such as response times or availability.

- But wait! Maybe the product can be simplified *under the surface*. Architectural compliance could be sacrificed. For example, information could be hard-coded instead of being configurable. New software could be developed as a monolithic block instead of creating a modular solution based on microservices and an enterprise service bus. All of this will only cause issues for future development anyway...

Whether these are explicit or implicit thoughts of a project manager, they reflect reality in numerous initiatives. A project manager will first compromise in invisible areas and on targets he or she is not rewarded for.

That is why a Chief Data Officer has the highest interest in avoiding such situations. Among all Architecture disciplines, Data Architecture faces the severest consequences if not adhered to, even more so if this becomes a regular pattern across multiple initiatives.

But how would you address this topic? Let me suggest three areas of activity:

(i) **Create awareness.**

No project exists in isolation. To make people understand that initiatives have an impact on the rest of the organization, you may tell the story of the "3+1 dimensions of a project": You do not only have to consider the three project-specific parameters **Costs**, **Time**, and **Scope** but also the corporate parameter **Architecture**. The lower the Architecture compliance, the bigger the impact on the rest of the organization.

(ii) **Ensure the involvement of Architecture.**

It must be mandatory to involve Architects in each phase of a project or in each scrum, respectively.

Across all initiatives, your organization needs to actively manage its technical debt. The responsibility to follow up on data debt is with the Data Office.

Where an organization works Agile, the product backlog must contain an Architecture backlog.

(iii) **Show the financial implications.**

It is not expedient to ask for architectural compliance based on a "principle of order" or any other valid yet abstract argument.

Instead, you'd enhance the perspective from that of an isolated project to its impact on the entire organization (shareholder value principle). You'd do so by forecasting the effect of architectural noncompliance and by adding a price tag to it.

You should definitely consider the **costs of complexity** here, especially by illustrating the impact of a rising number of Architecture violations on future changes to your organization's business:

- Costs will go up, as numerous areas have to be assessed, adapted, and tested, probably together.

- Time to market will increase, as your organization may not have enough specialists to address all necessary changes in parallel. This may also require additional interface workarounds.

- Increase in risk of errors, as the same new logic would have to be implemented correctly in a vast number of areas.

If you think of concrete use cases, you will realize that complexity does not only make your daily work difficult. It is also a hindrance to innovation, the lifeblood of successful organizations (Figure 19-7). While this is usually hard to quantify, its financial impact should not be underestimated.

These three steps should lead to a balanced approach to Architecture compliance.

Depending on the overall situation of the organization, the Board will always reserve the right to overrule Architecture standards. This is their good right, and it is absolutely acceptable for them to do so – if they do it consciously and in full transparency.

**Figure 19-7.** The cost of complexity

## A scoping approach

I guess you would now want to know how to find the optimal solution, somewhere between the happy flow and the 100 percent solution.

Instead of relying on gut feeling or 30 years of business experience alone, you require data, so that you can base your decision on facts. (Why doesn't this come as a surprise?)

Again, to get the full picture, you need to consider the "3+1 dimensions of a project," that is, the project-specific aspects as well as the organization-wide view.

For the latter, you need to know the future costs of Architecture noncompliance, as outlined earlier. For the former, the following questions require answers:

- Where can things go wrong in any concerned business process?

- What are the (direct and indirect) costs of failure (remember the 1-10-100 rule described in Chapter 1)?

- What does it cost for the initiative to cover those points – both in building an adequate solution and in daily business?

But how do you understand where things can go wrong? Data can help here, as well!

The following approaches can usually be combined:

(i) **User stories**

The need for improvement that had led to the initiative at hand should have provided you with an abundance of use cases and user stories. Each pain point that was discovered has the potential to form a user story – usually including an observed deviation from the happy flow.

(ii) **Previous process results**

Have a look at results from comparable processes in the past or from any new processes' predecessors: Even if processes change entirely, many parameters from earlier versions or from related processes in other areas may help forecast the expected exceptions and even their frequency.

(iii) **Simulations**

If your organization systematically records user patterns on the subprocess level (e.g., the click behavior of visitors of your website), you will be able to tell in detail how certain user groups react in certain situations – even in a different context.

Knowing the probabilities of every single aspect of your target process doesn't tell you the final exceptions and their probabilities right away. But they usually let you run simulations with a considerable number of test cases. Here, a random number generator can simulate daily life's variability, as recorded earlier.

Thank goodness, this kind of simulations does usually not require personalized data. That is why you can collect anonymized data about customer behavior on your website as well as about employee behavior in front of office computers.

Of course, all of this is an iterative process. Based on your findings, you'd design exception processes – but these processes need to be validated as well, and they come with new cases, the "deviations from the deviations."

Feel free to apply all of this to your own data processes first. Those processes are undoubtedly prone to deviations from the happy flow.

# Analyzing Data

*"And now tonight's business news:*
*irrelevant statistics are up 27.45%, but*
*meaningless figures fell 110%"*

**Figure 20-1.** Useless Analytics

© Martin Treder 2020
M. Treder, *The Chief Data Officer Management Handbook*,
https://doi.org/10.1007/978-1-4842-6115-6_20

# Preconditions of meaningful Analytics

Are precise algorithms and high-quality data all you need?

If you tend to say "yes," let me try to disappoint you...

Even if you have reliable data and if you use mathematically proven algorithms, the results may not be watertight.

Here are the two main challenges:

- The preconditions for the applicability of an algorithm
- Ambiguous interpretation of rules

Let me use the example of the well-known ANOVA (**an**alysis **of va**riance) from Applied Statistics. This method is used frequently to compare variations of different groups within a sample. I will not go in detail but point out a few potential pitfalls.

## Are all preconditions fulfilled?

In general, Data Scientists gratefully accept the provision of mathematical routines through statistics software.

This easy access, however, comes with the risk of those routines getting used thoughtlessly. Preconditions and adequacy often remain unvalidated.

Let's have a look at a typical example: A one-way analysis of variance. The three critical assumptions for the reliability of such an analysis of variance are

- Response variable residuals are (approximately) normally distributed.
- Variances of populations are equal ("homogeneity of variances").
- Responses for a given group are independent and identically distributed normal random variables.

Validating those assumptions is not straightforward in real life. As an example, you can determine the residuals based on your sample, but you cannot verify whether they follow a normal distribution. Furthermore, it is difficult to tell whether the variances of different samples are sufficiently equal.

# Can the investigator influence the outcome?

The last point leads to a second challenge: Despite the availability of a correctly determined sample and the preciseness of the mathematical formulas, many decisions are left to the investigator. This finding should surprise everybody who considers data science a mathematical, fact-based, undisputable discipline.

Bias comes in two phases: First, in the data preparation phase, then in the data analysis phase.

The data preparation phase, also known as Exploratory Data Analysis (EDA), is an indispensable step to detect data issues early, to exclude first errors, and so on. If done well, it will lead to more meaningful results of the analysis. But it is not a mathematically precisely described phase. Most decisions are left to the investigator. In other words, they are based on experience and gut feeling.

The subsequent data analysis phase, despite the availability of mathematical algorithms, leaves a lot of room to human bias as well.

I will use the ANOVA again, as an example of a scientific approach that leaves a considerable amount of decisions and estimates to the human being.

Let's have a look at a training text of Zurich University, after describing the execution of an ANOVA:

*"(...) The Levene test tests the hypotheses that different groups have the same variances. In case the Levene test is **not** significant, homogeneous variances can be assumed. If, however, the Levene test was significant, one of the preconditions of the ANOVA would be violated. The ANOVA is assumed to be fault tolerant in case of light violations. Violations are considered unproblematic notably in case of sufficiently big groups of more or less the same size.*

*If sizes of samples differ significantly, a strong violation of the homogeneity of the variances leads to a distortion of the F test. Alternatively, you could use the Brown Forsythe test or the Welch test which are adjusted F tests. (...)"*[1]

---

[1] Website of Universität Zürich, www.methodenberatung.uzh.ch/de/datenanalyse_spss/unterschiede/zentral/evarianz.html. Own translation. Original text: *Der Levene-Test prüft die Nullhypothese, dass die Varianzen der Gruppen sich nicht unterscheiden. Ist der Levene-Test **nicht** signifikant, so kann von homogenen Varianzen ausgegangen werden. Wäre der Levene-Test jedoch signifikant, so wäre eine der Grundvoraussetzungen der Varianzanalyse verletzt. Gegen leichte Verletzungen gilt die Varianzanalyse als robust; vor allem bei genügend grossen und etwa gleich grossen Gruppen sind Verletzungen nicht problematisch. Bei ungleich grossen Gruppen führt eine starke Verletzung der Varianzhomogenität zu einer Verzerrung des F-Tests. Alternativ können dann auf den Brown-Forsythe-Test oder den Welch-Test zurückgegriffen werden. Dabei handelt es sich um adjustierte F-Tests.(...)* (called July 18, 2019)

I have underlined all words that leave room for interpretation and bias – an astonishing number!

As an example, take the term "significant" from the preceding text. You may argue: "In the case of statistics, significant and not significant are very precisely determined, aren't they? Therefore, there is little to no room for misinterpretation."

Yes, indeed, the border between "significant" and "not significant" is usually precisely **defined**. This definition, however, is not a result of an objective determination of the best possible (let alone "correct") threshold.

No matter how you measure significance, it will usually be a continuous function. As a consequence, the degree of significance on both sides of the selected threshold is almost identical, if you stay close to it.[2]

In other words, "objective significance" does not exist. While there are always good reasons for the chosen separator between "significant" and "not significant," its final determination remains somewhat arbitrary: You have to draw a line somewhere.

In reality, imagine two random samples taken from the basic population. Not surprisingly, they lead to almost the same analytical result. Now, be conscious of the fact that "almost" may mean that one result is slightly below the threshold, while the other one is above.

How do you proceed in such a case? Will you resist the temptation to select the one sample that supports your original assumption (or your boss's expectation)? What if somebody else uses the other sample? You come to fundamentally different conclusions, based on very similar statistical results.

This topic is a typical case of a mathematical statement that usually remains unchallenged. I invite you to challenge such statements even if they come from statistical textbooks. At least, you should understand the underlying logic and decisions to limit the human bias.

## How about combining both challenges

Sometimes you find both problems combined: The preconditions are listed without explanations, and in a way that allows for biased approaches. A beautiful example can be found on BA-Support.com, again about the ANOVA:

---

[2]That means you can get the difference as small as you want by moving both values sufficiently closely toward the threshold. Mathematically, you'd say that no matter how small you set the difference $\Delta$ in significance $s(x)$ ($x$ being a real number), there will be a value $\varepsilon$, so that the difference between $s(\text{threshold} - \varepsilon)$ and $s(\text{threshold} + \varepsilon)$ is smaller than $\Delta$.

> *"This test can only use when a number of preconditions are in place. These are that all groups must either contain <u>more than 30 observations</u> or <u>be normally distributed</u> also that there must be <u>comparable</u> variances across groups.*
>
> *Furthermore, the variable describing the groups <u>should</u> be nominal-scaled while the dependent variable (variable we compare with, or just 'the other!') <u>should</u> be interval-scaled. In some textbooks, you will also be able to read that the groups <u>should</u> be similar in size, do <u>not have to worry</u> about in SPSS, as it has a procedure to correct for unequal sample sizes."* (BA-Support, 2019)

Again, the underlined words form an impressive number of cases open to human bias. I invite you to ponder on the following questions:

- Where does the threshold of "30 observations" come from?

- Is there really a good reason to draw a line exactly between 30 and 31 observations – everything below is "bad," and everything above is "good"?

- What does "comparable" mean? Is this decision left to the investigator, his/her experience and gut feeling?

- "Should" implies that something is recommended but not mandatory. So, what happens if you do *not* follow this advice?

- Is it wise to trust a piece of software without knowing which logic that software uses to make a decision?

# General limits of AI

Artificial Intelligence comes with an incredible number of opportunities. But we need to be aware of the limits and how to deal with them.

## Data sources

The best algorithms lead to wrong results if fed with bad quality data. This should be intuitively clear. However, it is still broadly ignored, as the main focus often lies in obtaining results. And information derived from bad data doesn't look worse than if it is based on high-quality data.

As a consequence, in many cases, it is only the following activities that are expected from data scientists when it comes to finding data:

- Searching the Web using keywords

- Doing screen scraping where data cannot be downloaded in file format

- Downloading files (if necessary, at second hand)
- Reusing files somebody else had downloaded earlier
- Deriving the meaning of columns, for example, from header names or content

This approach has led to typical patterns of problems that reduce the value of data:

- The age of the data is often unclear. It may have been accurate a while ago but has meanwhile become outdated. (And "a while ago" stands for an ever-decreasing time span!)

- The level of completeness is often unclear: Data may have been filtered, or collection may have stopped prematurely.

- Definitions of columns are often guessed: As most flat files or spreadsheets come without a proper explanation, data scientists are forced to derive the meaning, for example, from header names.

- Bias! It is usually unknown whether a table was created with a specific manipulative purpose in mind. Even if an organization that provides data has a good reputation, human beings are involved in assembling that data. And you cannot tell from the content itself: An honest "5" and a manipulated "5" look exactly the same in a data file.

- Another data scientist may have removed or merged columns that correlate strongly with others, possibly for good reasons in the context of that data scientist's specific question. But maybe the subtle differences would have been the interesting parts in the subsequent use of the same data to answer different questions.

In response, you may wish to add specific responsibilities to any data scientist's data acquisition job:

(i) **Know the source**

It is important to obtain knowledge about creators and providers of publicly available data. Without this knowledge, data is as useless as a positive product rating on the Internet without knowing whether the author is the seller or got paid for the rating.

Data search portals such as https://datasetsearch.
research.google.com (finally released in 2020) are
great. But they are not "the source," and they don't
give any guarantee of Data Quality.

## (ii) **Contact the provider**

In some cases, getting in contact with the original creators or providers of data found on the Internet helps
you understand their motives and methodologies.

In the long run, a good relationship may come with
access to additional data or even a regular, fruitful
exchange between two parties.

## (iii) **Understand the history of the data**

Data Quality is not a constant attribute. It changes
even if the data itself stays the same. Finding out about
a data source's history of Data Quality helps understand the status quo: Has the data been maintained?
Is it under permanent review? Or was it the result of
a one-off "fire-and-forget" effort, no matter how diligently it was executed?

## (iv) **Clarify the data model**

The data model and the terminology used by a data
source will probably differ from how your own organization calls data elements and relationships, and
how it defines data structures.

A bit of research at this point helps avoid severe trouble later or even undetected misunderstandings that
lead to wrong conclusions.

## (v) **Check multiple sources**

Hardly any information is only available through one
single source. Searching for multiple sources of the
same data is, therefore, a good principle, as it allows
you to validate through comparison.

Please consider that different parties often copy from
each other. Just as with rumors in your daily life (which
appear more plausible if you hear them from different
sources), this may result in a false sense of confirmed
quality.

### (vi) **Objectively classify the data sources**

In order to be able to reproducibly qualify the results of your data science initiatives, you need to classify the quality of each data source. This is also important for future use of the same data in a different context (and maybe by a different data scientist).

Typical attributes are reliability of content, the correctness of the description, and the frequency of updates.

When you determine the level of bias and trustworthiness of a data source, remember that it is not only about the collector of the data. The resulting data may not be based on facts but on a "representative" survey or on estimates.

A survey is a valid means of getting a good understanding of a situation in cases where a full assessment comes with unacceptable effort. But the interviewees might be biased.

If, for instance, you ask 1000 CEOs about their organizations' maturity level of digitalization, or if you ask 1,000,000 customers about their level of satisfaction, you have a comprehensive sample – but each answer will be biased, and the interviewees' criteria may differ.

### (vii) **Document your data sources**

Information about data sources (both about the data files themselves and where they come from) should be documented (and maintained) together with all other relevant metadata so that the knowledge is systematically made available to all data scientists.

The documentation should contain all known attributes such as the (confirmed!) meaning of columns, the covered moment in time or time period, the year of creation, any applied filter, and the original purpose of the creation of the file.

Furthermore, you may wish to standardize your criteria and attributes so that any two sources can be compared.

(viii) **Collaboration**

If your organization's policies allow, I encourage all data scientists to work with a broader data community, beyond your own organization. This is not about disclosing internal information but about sharing information about external data sources.

# AI algorithms

Let me illustrate some limitations of AI, using a concrete example.

This example is about face recognition, one of the most prominent use cases of neural networks.

I personally am impressed by what face recognition can already achieve. However, using such algorithms to pick suited candidates from a list of job applicants comes with a few challenges that may make you reconsider their suitability.

And I am not even talking about the ethical part – even the service itself cannot be worth the money spent!

In Autumn 2019, Bobby Hellard published an article called "AI and facial analysis used in job interviews for the 'first time.'" It describes a case of a US organization that developed a software "which analyses the tone of voice, vocabulary and facial expressions in video interviews to determine a candidate's suitability" (Hellard, 2019).

The algorithm may accidentally be biased. Even worse, it may be *biased on purpose*. The creators could shape it in a way that their own CVs are rated highest, making themselves look like the most auspicious candidates. In essence, AI allows people to create perfect job opportunities for their own career progression in a very unethical way. After all, it is a black box – nobody else will ever find out...

The worst aspect, as far as I am concerned: Such an approach prevents HR from getting better over time!

Here are a few shortcomings of the underlying AI algorithm. You will find them in other cases as well:

(i) **Manifesting errors of the past**

People with combinations of attributes that prevented their holders from being selected in the past will not be selected in future. If the selection criteria of the past were suboptimal, so will be the criteria of the future.

(ii) **Inability to consider changes in demand patterns**

The world changes and different profiles are required. How would such a backward-looking algorithm be able to understand changes in demand?

(iii) **Inability to judge the learning level**

Unsupervised learning is key to improvement. In this case, however, the success rate can hardly be validated, particularly not by the algorithm itself. This is not a problem of the algorithm but of the available training data. After all, it takes years until you can tell whether a new hire can be considered successful.

(iv) **Not knowing WHY**

As with many AI algorithms, the algorithm at hand does not tell us *why* specific profiles have been more successful in the past. To determine the best candidates, you'd need to base your hiring logic on meaningful parameters. An algorithm that cannot distinguish causality from correlation is useless and dangerous.

Don't get me wrong: We should not assume that the determination of correlation is always insufficient! But we need to be clear about when we need causality and in which cases we can work with correlation.

When it is about finding out whether we should "Do more of A to achieve more of B," causality is crucial. If, however, we want to validate a statement like "Look out for C to find more of D," working with correlations may be perfectly okay.

(v) **No insight beyond the training data**

A big challenge of AI is the applicability of results. No AI algorithm can provide reliable insight beyond its training data. If an element is "evaluated" by an algorithm without being covered by the algorithm's previous training or testing data, the result is useless at best. Such a situation may be acceptable where an algorithm tried to improve the average selection, for example, the number of holes you need to drill in the ground on average until you find particular raw material. You could consider it a success if 50 percent of all cases saw an improvement, while the other 50 percent were just random – the overall effort would go down. But whenever you are dealing with individuals, improving on average is considered ethically unacceptable in most modern societies.

Why do we need a data-literate HR world? To prevent approaches like this one from becoming part of an HR department's toolbox. No data manger or privacy officer should have to tell the HR people!

## (vi) Legal issues

Organizations that use this kind of algorithms usually stress that it is one evaluation component only and that human beings still do most of the assessment before a hiring decision.

This statement, however, does not apply to those CVs that were sorted out by the algorithm *before* the first human being had a look. They remain 100 percent processed by software, no matter how much human evaluation is subsequently applied to the other CVs.

That is why this approach is not only ethically questionable, but it is also violating existing laws. Article 22 of GDPR states "The data subject shall have the right not to be subject to a decision based solely on automated processing."

All in all, you don't want to lose control over the process (as illustrated in Figure 20-2). The best advice I have read in this context comes from Hanover Recruitment Limited, a British recruitment agency. They state on their website: "Ultimately, you should treat candidate and client data the way you would want your own data to be treated!" (Beatie, 2018).

*"Actually, yes, we did let AI choose
the shortlist of candidates!..."*

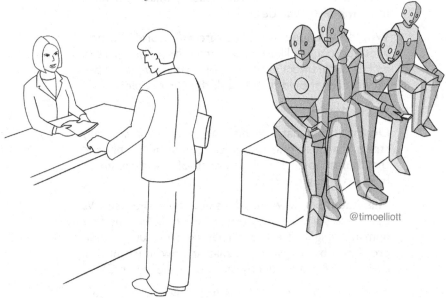

@timoelliott

**Figure 20-2.** HR = "Humanoid Resources"?

## Human behavior

"I know what I am doing" is what I often hear from Data Scientists. But it is not sufficient for the scientist to *know*.

Data Science is about credibility. It is a matter of time until two colleagues independently work on the same problem. They will probably use different (variations of) methods. Please expect them to interpret the same situation differently and to come to different results.

These results may not be fundamentally different. But they will at least differ sufficiently even for a layperson from Marketing to question the preciseness suggested by the number of digits after the decimal point.

What helps? **Transparency!** Share what you are doing, in layman's terms. Don't overpromise – you don't need to! Explain limitations. It doesn't hurt to say, "This result is only valid if A and B are independent, but we don't know to which extent they are."

# AI – Quo Vadis?

The game Go is reported to be one of the most complex games on earth. You will have read that, in 2016, DeepMind's AlphaGo computer defeated South Korea's Lee Sedol, the reigning world champion. This event was broadly perceived as a massive step in the development of Artificial Intelligence (AI).

I recently read an article from IBM that claimed: "The important outcome from Sedol's defeat is not that DeepMind's AI can learn to conquer Go, but that by extension it can learn to conquer anything easier than Go – which amounts to a vast number of things."

Sounds reasonable? Well, it may not be.

The main problem I have with this statement is the word "easier." Its comparative degree suggests that we are talking about a one-dimensional scale. In other words, you could conclude that any two intelligence-based performances can be sorted by "ease" (at least if one of the two is "playing Go").

But how do you define "easy"? Is it *easier* to defeat the reigning world champion in Go, or is it easier to convince a kidnapper to release all captives and to give up? Is it easier to win *Jeopardy!* or to convince someone of a different view during a discussion? We obviously face multiple dimensions of human brain performance here.

Ask a computer like AlphaGo to calculate the best next movements of a football player in real time, and (net of hardware constraints) even the author of these lines would turn out to be superior (which says a lot). Whoever has ever watched RoboCup – the world championship in robotics football – knows what I mean…

I don't know how quickly AI is going to progress. However, one indicator is the investment of money in research. My recent discussions with a lot of organizations suggest that they are moving from the phase of "We need to be part of it by all means" to that of "Show me the value." The real money will go to AI research that comes with a business case.[3]

So, if you put yourself in the shoes of an organization, how would you think and act?

A well-managed organization will inevitably ask "How can AI make us more successful?", considering the usual stakeholder triarchy of customers, shareholders, and employees. Even universities and nonprofit organizations in general are more and more driven by the need to work toward creating tangible value.

---

[3]Please don't focus too much on those public relations–driven "Innovation Hubs" of big corporations where researchers can play around without commercial pressure as long as they publish interesting stories from time to time.

Let us be honest: It will take a huge, long-time investment to have true Artificial Intelligence (in the sense of copying the human brain) add value to our society (or even to single organizations). So, what drives the investments of organizations?

Firstly, organizations generally prefer dedicated specialist solutions for each task. This is not specific to AI. No car manufacturer would want to invest in a vehicle that can transport bulky goods, win a Formula 1 race, and deliver a luxurious convertible feeling at the same time. Three specialized solutions will usually do a better job in their specific areas, at lower overall costs. The fact that you need three solutions for three different tasks is not perceived as something embarrassing, for good reasons.

Look at robotics: While the media is full of humanoid robots that can smile and shake hands, most commercially driven research is still being done on specialized robots which can do precisely one thing in a near-perfect way (and which requires at least substantial reconfiguration before being able to perform another task). The industry has been working on this for decades, with impressive success.

Secondly, developing something further that is extremely far behind existing capabilities (in this case, current human abilities) has a bad business case. Return on investment lies in the distant future, and it is uncertain, after all.

That is why I am reasonably confident that development in AI will continue to focus on areas where AI already delivers a performance superior (or comparable) to that of humans today, through specialized solutions. Mimicking ingenious mechanisms of the human brain, such as a neuron's way of working, will definitely contribute to the success of this approach.

At the same time, at least in the foreseeable future, I don't expect any substantial progress in developing clones of human brains that don't require reprogramming for new challenges.

Does this sound pessimistic? It is not meant to!

I am a great believer in AI's growing capabilities in *complementing* the abilities of human beings. God did not design us godlike (which is easy to believe after watching the evening news), so why should we, in turn, invest our precious brainpower in developing humanlike devices?

# Recommendations around Analytics

Based on what I have said so far, I would like to propose a set of guidelines that you may ask your entire Analytics team to follow: My "12 Guidelines for Analytics."

You should feel free to adjust them to your individual findings and priorities – but I strongly recommend having such guidelines as they help achieve a consistent approach across the entire team.

# I. Determine the necessary degree of preciseness

Sometimes it is not required to consider all mathematical preconditions.

This is particularly the case if it is only about finding a preoptimized starting point for a second algorithm (through heuristics). In this case, the quality of the starting point has an influence on the amount of the remaining calculation work but not on the final result.

Sometimes, however, convergence is not guaranteed, and often you cannot validate later whether your algorithm plus data worked well. Even if you split your data into training and testing data, the latter may not correctly reflect reality.

But wherever your model is "only" expected to make a process more efficient, where you have a short validation cycle, and where you can measure success easily, you should not insist on a minimum level of accuracy up front.

It doesn't matter whether you have found the best possible algorithm – if it makes things better, its existence is justified. If a process performance was at 70 percent, and you get it to 80 percent, it adds value, despite the fact that 80 percent is far less than 100 percent.

In such a case, you can operationalize a model within days – after you have run it in parallel, and the resulting algorithm delivers improved results.

And once your solution is live, you are in an even better position to further fine-tune it, particularly if you can measure the delta between its current performance and 100 percent.

Putting such progress into charts helps impress your business customers: You can link something your team has done immediately to an improvement of a metric that is relevant to a business stakeholder.

# II. Don't use a formula just because "it works"

Let's assume you have an interesting case. The question is clear; the base population is already available in your database. You can call a function of your statistics software, and it will return a syntactically (and mathematically) correct result. All input parameters are available. The temptation is high.

But does the formula fit the purpose? Maybe there is a better approach – perhaps even using available business information?

Example? You might have used an Unsupervised Learning routine to cluster your base population. The number of clusters follows scientifically accepted criteria, for example, using the sum of squared distances within each group. Your gut feeling tells you that you have found a good compromise – increasing the number of clusters further would not add significantly to the score you have defined, also taking into consideration the reduction in cluster size that comes with any further clustering. Let me say it in economics language: Your marginal benefit in increasing the number of clusters approximates zero.

So, whatever number of clusters you have finally decided to go for, your decision is based on a blend of data, formulas, gut feeling, and experience. However, it may **not** have been based on the business background of the underlying population.

This is why data scientists need to be ready to get out of their comfort zone. There is a lot of complementing wisdom on the business side. Discussions with the right business folks may reveal, for instance, that a natural clustering of your base population is already evident from a business perspective, using combinations of well-known attributes.

Using business knowledge may not only be more adequate than applying a broadly accepted yet content-independent method such as an elbow criterion (where you look out for a strong bend along the benefit curve, indicating a significant drop of the marginal gain). It will also make it easier for business folks to understand what you are doing, and your credibility as a practical solver of problems will increase.

## III. Check all preconditions

Is your basic population normal-distributed? Really? Can you guarantee full statistical independence?

You cannot tell from the result whether all preconditions of applying a method were met. And where it is almost impossible to fulfil a precondition, you should at least know to what extent it is fulfilled.

Wherever possible, you should schedule regular validation of whether all preconditions were given. In some cases, you might find that your results were too bad for the preconditions to be true – and sometimes your results corroborate your assumption that all preconditions were fulfilled.

Think of election polls where people coming out of their election cabins are asked which party they have just voted for. The result is then displayed as the first prediction after the polls close. The institute usually selects the polling stations in a way that their joint result of the last election was very close to the outcome. Furthermore, those people need to be randomly selected, and all of them need to voluntarily tell the truth.

You will remember that the results of such polls are usually relatively close to the final outcome of the election. But are they as good as they should be if all the statistical preconditions were fulfilled? The answer is NO. If you do a bit of maths, you will find that, statistically, 98 percent of all polls with fulfilled preconditions would be better than the observed result!

Of course, not all your calculations in a business context can be validated as quickly as this kind of election polls. But if you have the chance to, say, check your forecast of the sales distribution across the product portfolio against the real sales figures later, you should do so, to get a better feeling of the validity of the preconditions.[4]

## IV. Be open about the limitations

Both data and algorithms have natural limitations. You might have reasons to take the risk of using them beyond those limitations. But be clear about your choice and motives.

Sometimes you may be forced to perform year-on-year comparisons where the annual data comes from two different sources (e.g., because new software was introduced in between or because the data provider has changed). Please add this information to the presentation of the results, as a warning.

Another typical limitation is the origin of the training data for your AI models. You usually cannot assemble it yourself, particularly if you want to base your analyses on millions of records. As a consequence, you usually end up using the same limited number of publicly available data repositories. Here's the issue: The fact that your results are in line with those of others does not prove their correctness. It may rather indicate that all of you have been using the same source of data.

Look at the repositories of images for OCR and image recognition: They have usually been assembled in a specific context, and they will probably not work beyond that very context.

There is another challenge that comes with character recognition: No OCR algorithm can safely tell the difference between a 0 (zero) and the capital letter "O" or the difference between the "1" (i.e., the numeral "1" written in Anglo-Saxon style) and the capital letter "I" (if written without serifs). The same image may mean two different things. Even within the family of digits, a European "1" and an Anglo-Saxon "7" may look identical. For a safe distinction, you will also need to consider the context. If you cannot do this, be open about it!

---

[4]This doesn't mean that you have to tell the Board that you worked with unfulfilled preconditions.

## V. Explain your assumptions

Please have another look at the clustering example earlier and remember the analysis of variance at the beginning of this chapter. How often did you make a human choice?

- Defining a score
- Choosing the metrics of how similarly the elements of a cluster behave
- Balancing the variance within each cluster against the effort that goes up with the number of clusters and so on.

We are far away from the necessary level of unsupervised learning that would allow us to leave all of this to an algorithm that knows autonomously which information to take into consideration. That means it is okay for humans to make decisions.

But you need to explain those decisions. Which assumptions have made you decide the way you decided?

## VI. Don't convey a false impression of preciseness

Imagine someone collecting the passports of a team of 40, recording their ages, and calculating the team's average age. Let's assume the resulting average age to be 34 years, rounded.

Now imagine another person ignoring the passports and estimating the age of each team member instead. That person calculates the average age of the team to be 37.575 years.

Which of the two persons provides a more accurate result?

I guess you see what I am aiming at.

Being open about the preciseness of your results is part of building your credibility. Instead of "the likelihood of a customer in this group to repay a loan is 68 percent," you may wish to state:

"With a probability of 95 percent, the likelihood lies between 65 and 71 percent, assuming that all preconditions of the algorithm are met. In other words, if the true likelihood of that group were 68 percent, on average 19 out of 20 simulations would forecast a likelihood of between 65 and 71 percent. For each precondition that is not fulfilled, the uncertainty increases."

This statement is harder to digest, but once people get it, you won't need to repeat it over and over again. You'd achieve the following targets:

- You support your organization on its journey toward a data-literate organization. They don't need to be able to prove the Bayes theorem, but they should gain a basic understanding of what is possible.

- You manage expectations and protect yourself: People should have a realistic idea of how precise your figures are. And you don't want to be "proven wrong."

You always leave room for improvement. More data and more performant computers will generally increase accuracy. This allows you to create a price tag for increased accuracy. If an executive is not happy with a range of six percentage points as in the preceding example, you know what your proposal will have to look like.

# VII. Automate data preparation carefully

Automation of data preparation can save a lot of time, and it leads to reproducibility as it reduces the human factor, including personal bias.

Note that this does not guarantee (or even improve) the correctness of the data. Beyond reducing human error and speeding up the process, it just makes the same data look better, and it eases subsequent processing. This, however, may foster a false impression of Data Quality.

# VIII. Use DataOps

With the rise of DevOps, that is, having the same teams develop and run a piece of software, people started to think of extending this idea to data topics. Unsurprisingly, this concept got coined DataOps.

DataOps helps set the incentivization properly by avoiding a situation where different parties work toward their own targets only.

Furthermore, the review cycles and the constant monitoring of Data Quality help discover implausible data, for example, through heuristics.

But please resist the temptation to limit this approach to Analytics. Starting DataOps with the preparation of previously unmonitored data for Analytics purposes is too late.

# IX. Balance diligently

You will hardly ever find the overall optimum for a problem. The world is *simply too complex* (pun intended). There are too many influencing parameters, and the different optimization criteria can usually not be weighed objectively.

This situation forces you to compromise – either by further simplifying your models (every model simplifies by definition) or by reducing the list of options to assess (as you easily have more options than there are atoms in the known universe).

You can apply both strategies at multiple levels.

- You can deal with simplification by building a model as complex as technically possible, or you can simplify as much as possible while still obtaining a syntactically valid result.

- Alternatively, you can reduce the number of options through clever usage of algorithms (e.g., I don't need to assess an entire group of options as soon as I have found one option that is proven to be better than each of the options in that group).

- Furthermore, you can also reduce your list of options before you apply an algorithm, based on nonscientific factors such as your business strategy. For example, you have determined a group of options, and you can prove that the overall optimum is part of that group. But none of these options is in line with what your organization has decided to focus on as part of its product strategy.

But how do you know to which extent you should compromise? Again, you would have to calculate. And, again, you would quickly find out that the determination of the "best" level of compromise is a tremendously complex calculation, far too complex to be executable.

The first challenge would be the **measurement** of how far you compromise. Insisting on fully considering **all** parameters would obviously mean zero percent compromise, while a random choice would be 100 percent compromise. But how do you find out where in between any approach would be?

Complexity cannot be overcome by developing more and more sophisticated algorithms. The more sophisticated an algorithm is, the more complex it gets. And the more complex an algorithm is, the less you know how accurate the result is. In other words, you don't win anything compared to guessing the right balance.

We often think we are forced to go straight from old-fashioned gut feeling to modern data science, and that data science is going to replace gut feeling. This way of thinking suggests that data science is the solution and that it will provide the ultimate answers.

Instead, the best choice may remain a well-balanced approach somewhere in between the two extremes. This, however, might require a change of ambition: From aiming at finding out what "the right business decisions" are toward determining how to apply data-driven approaches in the *most adequate* (not perfect) way.

And, yes, experience and gut feeling remain part of this exercise – which stresses the importance of being open about the limitations (see Guideline IV "Be open about the limitations")

## X. Exclude emotional factors

Another helpful skill is your ability to distinguish between "rational" and "emotional" decision factors. As we have seen, even the rational factors are deficient. That is why you should at least eliminate emotional factors such as pride or anger.

How do you develop such skills? You can, of course, read books about Emotional Intelligence (and I encourage you to do so), but unlike studying science, you cannot just "read and remember."

That is why I recommend that you reflect together with others. Be open-minded and expect others to surprise you. Again, unlike science, you will find that two different views may both be acceptable.

Therefore, this approach does not mean adding another piece of knowledge or a rules engine to some knowledge store in your brain – you might rather wish to broaden your basis for future decision-making.

## XI. Consider changes outside the model

In traditional modeling, you consciously decide which parameters you want to become part of your model. In Macroeconomics, this is often described as Ceteris Paribus (Latin for "everything else equal"): Whatever you do not want to (or cannot) build into your model is assumed to be constant. Consequently, you are forced to think up front about **all** aspects that have a potential impact so that you can consciously decide whether to make them part of your model.

In neural networks, you often rely on the model to do this work for you. It is expected to implicitly discover all relevant aspects: If these are part of the training data, they will automatically influence the training process positively.

The missing need to consider *all* aspects up front comes with a risk, though. It is about the representativeness of the sample data you use to train your model and to validate it afterward.

In fact, you may miss certain aspects when collecting that data. You think your data is representative, but you may have accidentally left out relevant subsets of your overall population.

The most frequent omission is not to consider the **time** aspect. As a matter of course, any sample was gathered in the past. Well, things may have changed since! The COVID-19 pandemic with its dramatic change in consumer preferences is a good example.

There may be changes to aspects you would not even have thought of, but which may have an impact on what you are assessing. As another example, changes to legislation or scandals made public may have a significant impact on people's buying preferences as well, so that a sample from before such an incident would not reflect today's reality.

The issue is that this kind of failure is difficult to spot. You would, as usual, use 70 percent of the sample to train your model and the remaining 30 percent to test your algorithm. As a result, your test data would confirm your model as it reflects the same time in the past as your training data.

The good thing is that you can do something about it. The downside is that this comes with additional effort: You'd have to think about all potentially impacting aspects, even those far away from your core considerations.

It is, in particular, not sufficient to consider all attributes you find in your sample, no matter how reliable that sample may be. You are back to the hard work of deterministic models, to a certain extent: Diligently contemplating about all potential influence factors.

Of course, you can take some mitigative action as well: Use the longest possible time range for your training data and separate it by time period. If you model works similarly with data from different time periods in the past, it seems at least invariant to *regular* changes (such as fashion trends or the change to a different government) or to the presence of *singular* events (such as a pandemic, a natural disaster or the Olympic Games). It won't help in cases, though, where all time periods covered by your training data are before an impactful day X while you try to make forecasts about the time after day X.

Here is my key learning: A good model requires more than a perfectly trained Data Scientist – a broad educational background is almost equally important. This is a strong reason for mixed teams where different skill sets complement each other.

# XII. Define success comprehensively

How do you validate whether an AI initiative was successful? Or, in case you are improving a model iteratively, how do you know you are heading in the right direction?

At first glance, it is easy if a measurable business criterion was defined and measured up front. You see improvement, and you can prove that your algorithm was successful, even if you don't know whether you could have done better.

This judgement may be wrong, and the reason could be an incomplete measurement of the impact beyond the single target criterion you had concentrated on.

Imagine a police station that wants to become proactive in crime avoidance. We can assume that there is enough historical data for them to develop and train an algorithm to determine any (known) person's probability of committing a crime soon. As a consequence, the police station could implement a policy to put people to jail whose calculated risk of committing a crime exceeds a certain threshold.

Of course, even the most autocratic countries would probably not implement such a concept. But the underlying principle may sound familiar to you as it keeps being applied to other use cases.

Here's the interesting aspect: In the case described earlier, you would most definitely find that the crime rate will indeed go down! The AI algorithm will identify a lot of truly dangerous persons who would not be able to commit a crime while in jail. The initiative is going to be considered a success, in the name of safety.

But aren't there a few undesired side effects?

If you don't consider the entire impact of an initiative in your evaluation, you may miss the consideration of key factors (basic human rights in this case) that can render the entire result negative.

You may not find equally drastic examples within your organization. But you will often discover a similar pattern.

A very simple example is the assessment of different measures to increase revenue. If one of the measures is to apply discounts, you would not just validate the increase in revenue (which is almost certain in this case). Instead, you would also consider the negative impact of the discounts on the margin (and eventually on your organization's bottom line).

Here's my recommendation: Determine all areas that are possibly impacted by a data-driven change and consider them in an overall target function up front.

Such a weighted target function also forces the business stakeholders to make up their mind and to set priorities.

Admittedly, translating every parameter to one denominator (most frequently the financial impact on an organization's bottom line) requires a lot of work. Alas, data can help you do this job, and you are in Data Management!

# Explainable AI (XAI)

Unlike in the past, where algorithms were built from rules, many modern AI methods are based on learning (thus the expression *Machine Learning*), either from examples or through rewards.

The traditional, **confirmative** approach, as known from Operations Research and matured during the second half of the twentieth century, always starts with the determination of all rules and constraints. These are then translated into a model, to which a suited (mathematically proven) algorithm is applied.

This approach comes with two main risks:

(i) **Openness to bias and fraud**

The confirmative approach starts with a set of rules that are sitting outside the precise algorithm. They are usually assembled by people with a vested interest in the underlying issue who may therefore be preoccupied or even working toward the desired outcome.

As a result, boundaries are omitted, constraints underestimated, or inequations "tuned." Another significant source of bias is the creation of the target function, that is, the weighting of the different target parameters for the overall normalized target.

The resulting "problem," usually an inequation system,[5] can then be calculated with mathematical accuracy – which suggests that the result is indisputably correct. However, if the underlying model has been subject to bias, the result will rather reflect the ambitions of the investigators than an objectively right optimum.

---

[5]Such an inequation system is usually represented by a matrix or by more efficient notations in the case of network optimization tasks where the matrix would be extremely sparse.

(ii) **The incompleteness of the model**

Each optimization problem has an infinite number of parameters. Any list of nodes, boundaries, and constraints represents a simplification of the true problem.

While in the early days computing power was the main limiting factor, the primary constraint today is the completeness of the underlying business knowledge. If people don't know all constraints, they cannot add them to the model.

Of course, models are simplifications of reality. However, in Operations Research, you cannot tell up front whether the model is sufficiently representative of the real situation, and the result will not tell you either.

Only if you use the outcome in the real world, you may find out how good the model was. In most cases, however, you will end up with an invalid situation as certain constraints turn out to have been missing in the model. This situation requires a long iterative process that cannot be left entirely to computers.

This situation suggested that the approach be turned around, and to give the parallel development of Artificial Intelligence a second chance: You start with data, followed by the derivation of correlations and causalities.

This transition from rules-based approaches to neural networks means a change in direction:

From

**wisdom-to-data** (confirmative)

to

**data-to-wisdom** (exploratory/Bayesian)

You may have guessed that this move has not solved all the problems of mankind. In fact, it has even brought new issues.

Let's have a look at some striking challenges.

# Unknown cause and effect

As we have seen, today's algorithms work as a black box: You don't know anymore why a neural network comes to a result. *Causality is unclear.* Neural Networks are all about nodes activated and firing, based on thresholds…

How do you improve an algorithm of which you don't know how it takes decisions? It is not possible to fine-tune the rules – there are no rules, after all, at least not in a deterministic sense. The only way to improve a neuron network's quality of decisions is through more and more training. But what if you run out of training data?

And even when the known cases you have kept for validation purposes confirm a sufficiently high success rate, how about false positives?

Imagine an algorithm that successfully recognizes 99 percent of all images with a dog. It may even identify the type of animal on the other images.

But how do you know that, in real life, it will not recognize an (untested) arbitrary color pattern as a dog? The threshold between "It probably is a dog" and "The probability of it being a dog is higher than the probability of it being any other animal" is usually determined through manually set thresholds.

I am almost certain that science will understand these black boxes better over time. But it will be a long and tedious discovery journey, comparable with the decades-long (and still ongoing) process of understanding the human brain itself.

## Trust issues

It is usually okay for a patient to know that a pattern recognition algorithm supports the doctor in detecting skin cancer. After all, the results cannot be worse than human judgment alone.

But how about algorithms that may take disadvantageous decisions for single humans?

An AI algorithm doesn't know anything about the subjects of its calculations. It usually cannot say "Wait, this person filmed by a surveillance camera cannot be Ronald Reagan – it must be a carnival mask!" Instead, we'd need to enrich or modify the data up front, as the algorithm itself is not rule based.

You will easily be able to think of more practical examples with relevance to you and me, such as AI algorithms that calculate credit scores.

Another example of such a dilemma is an algorithm in autonomous driving that may have to decide between two options, both expected to cause casualties. Would you let an algorithm steer your car that may come to the conclusion that your life is of less value than that of a group of people on the road?

# Ethical issues

Evidence-based algorithms don't have a conscience – whereas rules-based algorithms can avoid unethical results by making such outcomes unattractive through the intelligent setting of boundaries or by selecting different weights in the target function.

But how do you find out that an outcome that an AI algorithm considers optimal is, in fact, an unethical outcome?

Remember the HR example under "General Limits of AI" earlier in this chapter: Here we would systematically sort out candidates based on an algorithm of which even the Data Scientists don't understand why it decides the way it does.

This is not only a violation of GDPR and other data privacy laws, it also discriminates against certain groups of people, *without* the developers' intent to do so! In other words, it is not sufficient to *feel* ethical. A Data Scientist also needs to *act* ethically, by consciously searching for factors that may have an ethically undesired impact.

# Is there a way out?

This question is being asked increasingly frequently, and it has led to the phrase "Explainable AI" (XAI): *Understanding what is going on* is considered critical to addressing the issues of **black-box algorithms**, of **ethics**, and of **trust**.

This discussion has only started, and there is no satisfying answer yet.

However, the following questions may guide us through the upcoming dialogue:

- How are Data Scientists going to gain the trust of users? Nobody trusts in an algorithm anymore that selects candidates based on experience which may be suboptimal.

- In which cases will users accept the "black box," and when will they refuse it?

- And what to do about it? Will there be new algorithms that allow for decisions to be tracked back? To which extent will it be possible to derive "rules of life" from the gradient of a learning AI algorithm? Will enhancements to known algorithms such as the recently promoted "Layer-wise Relevance Propagation" (LRP) mature sufficiently to add transparency to existing neural networks? Or will new approaches dominate where transparency comes by design? You might, for instance, start with subproblems with a lower number of dimensions, where a lower number of aspects can be understood more quickly, and where humans might still be able to understand the smaller number of interdependencies?

- Will it be possible to apply another algorithm to the output of such a "black box" that shows which parameters (or combination of parameters) have led to a yes/no decision?

- If transparency approaches develop, how will it be possible to classify them, so that data scientists will know which algorithm they are allowed to apply in which case?

- Do you think recent initiatives to shape Explainable AI will be successful any time soon? Or will people get used to trusting AI, as they "learned" to trust their IP television device or their Alexa device?

- Where else in our business lives do we face black boxes? Where have we always accepted them as a matter of course, for example, because organizations would not want to be forced to reveal their business secrets?

- If all Data Scientists use the same small number of analytics libraries and sets of training data, will we accept the higher impact if one of these is biased?

- Will we end up having ethical standards that will primarily depend on the impact of an algorithm? Will data scientists be able (or even be obliged) to check the law before selecting or developing the best-suited algorithm for a given problem? Will organizations develop ethical guidelines ("Responsible AI Policy") to gain trust beyond what the laws demand?

- How complex will it be for an individual Data Scientist to comply with all of these "nontechnical" guidelines? How complex will it be for a Head of Data Analytics to ensure the entire team behaves in a compliant way? Will it be possible to audit compliance?

This discussion is not based on the assumption that we can make all existing AI algorithms sufficiently transparent. But it will certainly become possible for some of these algorithms, and other algorithms may allow for partial transparency. Imagine, for instance, a two-step algorithm that uses the first phase to develop a set of rules based on training with test data. The second phase would then apply the most appropriate rule, which is possible in near real time as the complex calculation happened during the first phase. Such an algorithm can share which rules are selected and applied during the second phase.

You still don't know **why** these rules were found to be the most appropriate ones in a particular case, but at least you can check for potential flaws where a rule turns out to be objectively wrong or discriminating.

Finally, a trade-off between effectiveness and "explainability" of algorithms may become both possible and necessary. AI solution providers may be forced to reduce the effectiveness of algorithms in order to comply with laws that ask for a minimum degree of transparency.

Such regulation can be expected to come from the European Union, as can be read from their White Paper on Artificial Intelligence, published in February 2020 (European Commission, 2020). Here, transparency has been determined as one of the Commission's seven key requirements of their regulatory framework for AI.

On page 15 of their White Paper, the Commission states:

*The lack of transparency (opaqueness of AI) makes it difficult to identify and prove possible breaches of laws, including legal provisions that protect fundamental rights, attribute liability and meet the conditions to claim compensation. Therefore, in order to ensure an effective application and enforcement, it may be necessary to adjust or clarify existing legislation in certain areas [...]*

It is advisable to follow the development carefully, in case your organization falls under EU legislation or wants to be active in any of the EU countries. For ethical reasons, you should consider doing so no matter which situation you are in!

# Data Management in Crises

"Needless panic!"        "Inadequate precautions!"

@TimoElliott

**Figure 21-1.** Preparing for crises

© Martin Treder 2020
M. Treder, *The Chief Data Officer Management Handbook*,
https://doi.org/10.1007/978-1-4842-6115-6_21

# Prepare for the crisis

## Think the unimaginable

In one of my data roles, my team and I had prepared for any crisis we could imagine. Our computers were full of great tools and data. Then a ransomware attack hit the entire organization. Our company had not been the target, but the ransomware spread into our network. All of our Windows-based systems got infected, and the network went down within minutes.

A team of gifted data scientists, data analysts, and project managers ended up sitting behind their desks, without a single PC or notebook. Of course, all of our crisis plans saw us addressing the problem by utilizing our computers.

Similarly, until 2019, organizations around the world had great emergency plans, all of which started with people coming together to jointly evaluate the situation and form action groups. The COVID-19 crisis that began in early 2020 required such action groups, but the fight against that very crisis suddenly made physical meetings impossible.

When you prepare for a crisis, learn from previous crises but don't stop there. Imagine the unimaginable!

You may not have to go as far as preparing for an event in space that emits intense radio waves, which makes telecommunication on earth impossible, and... well, why not?

You may not have to imagine *all* the possible reasons for crises, of course. But how about turning it around and listing all indispensable aspects of your organization's business? After all, the applicability of an emergency plan usually depends on the impact, not on the cause of a crisis.

Ask your team to think out of the box, to generalize past crises, go through several processes, and determine all aspects required to execute the process.

You should include elements that everybody takes for granted – you consider the probabilities later in the process. Ask "What would be the impact if aspect X was not available?" Document all thoughts.

In a subsequent round, you can focus on data-related aspects. Where would data be part of the problem, and where could it be part of the solution?

Finally, think about probabilities. Most of the preparatory effort should be spent on scenarios with the highest product of (nonfatal) impact and probability.

# Be ready to prioritize activities

Have a look at what the members of the Data Office are doing today. All of these activities are useful – but which ones are essential? Prepare to stop activities that are "only useful" during a crisis, thus generating bandwidth for crisis-specific activities.

Any prework you do during normal times helps you save precious time in case of an emergency. And COVID-19 may give you a unique justification to prepare for crises next to your normal responsibilities.

As times change and employees come and go, you may wish to review the plan with your team from time to time.

# Be part of the organization's crisis plan

Any reaction to a crisis requires data. Facing a crisis, people may fall back into old habits of trying to find the data themselves, or they may even revert to gut feeling as their primary advisor.

It is, therefore, crucial that you prepare the organization to rely on data in such cases. The role of the Data Office must be institutionalized up front – trying to do so during a crisis may be too late.

This is why you need to talk to other leaders about crisis management as part of your regular communication.

Furthermore, you may wish to ensure the Data Office is involved in any corporate precrisis planning and that it automatically becomes part of the crisis response team. You do not want to waste energy at the beginning of a crisis on trying to get heard. Most disasters are of cross-functional relevance, and so is Data Management – a good reason for data to be in the center of any emergency planning.

Who from the Data Office should be appointed the "Crisis Lead"? The answer is straightforward: If a crisis does not justify the CDO in person to be in the lead, it is not a real crisis.

# Master the crisis

## Align with company priorities

All priorities change in a crisis, and they may keep changing as the crisis progresses. For you to set the right data priorities, it is helpful to align with the agreed Board priorities for the crisis. In other words, you would execute a "fast-track" version of the strategy alignment (see Chapter 4). This approach does not only ensure you are supporting the right targets. It also helps defend your activities against critics.

You should, of course, use the opportunity of being a member of the crisis response team and give input to the prioritization process.

## Don't try to become a hero

You will not be successful if you try to win fame. Fame should be considered a possible result, not a target.

When we faced the ransomware attack back then, the biggest risk was the loss of revenue, as it would dry out our cash flow. As a consequence, everybody was trying to ensure all rendered services could actually be billed to the customers.

Solving this challenge could make you famous. But we didn't see the need to be the tenth team dealing with this topic, and we didn't want to tell the others to let us data folks take over.

Instead, we observed the situation closely. We realized soon that the main risk was the opposite of revenue loss: Different groups had recovered data about transactions independently, which resulted in the risk of billing one rendered service multiple times.

We decided to establish an analysis of matching transaction records. We would hardly ever find two data records describing the same transaction at 100 percent certainty – but we didn't need to. The calculation of probability scores across millions of transactions pointed us to areas of systematic creation of duplicates.

Our duplication discovery added value to the organization, together with all the other teams' efforts to recover transactions. Eventually, all of us got praised together. If nobody had addressed the double-billing risk, though, nobody would have been praised, despite the enormous efforts by all involved parties.

## Prepare your team

Your team must not perceive the situation as "somebody's crisis" but as "our crisis." This is essential for your ability to act. That is why, as far and as quickly as possible, you should share this message with everybody from your team, ideally in an initial crisis meeting (be it face-to-face or remote).

You may convey the following messages:

- Clarify that each member of the team is more important than the organization. Safety and stability have the highest priority.

- Share the seriousness of the situation for the organization. Play with open cards.

- Stress that this team can do something about the situation: We may not be able to end the crisis or to solve all resulting issues, but we will at least be able to mitigate. And this may be essential for the organization to survive!

## Listen to your team

The first question during a crisis is not *how* to do the right things. It is: *What are the right things to do?*

The answer for the Data Office does not necessarily have to come from the CDO alone. There will most probably be time and room for brainstorming – use it! And even if you already have ideas, don't bring them to the table too soon. And be ready to discard your thoughts as you learn during the brainstorming.

## Structure your action

As with all changes, there will be a need for new activities, and certain existing activities will become more critical. At the same time, facing limited resources, other tasks need to get reduced or even stopped. Among the former, there will be activities to keep the business running, activities to measure the impact, and activities to suggest tactical steps.

You should always have a comprehensive bulleted list of all Data Office activities at hand so that a crisis response team or even the Board can set priorities (or, preferably, confirm your proposals), without running the risk for forgetting something essential. This step will help you stop providing some of the data services people got used to, but which are less relevant during a crisis.

Organizationally, crisis-related coordination activities should be consolidated into a crisis core team within the Data Office. This team would also set up direct communication channels with other entities to speed up response time.

## Manage the state of emergency

Depending on the magnitude of a crisis, some of the company rules and principles may need to be lifted or amended temporarily. Whenever you see the need to do so in the area of data, get the approval from the crisis response team.

Sometimes you may have to ask for a carte blanche so that you don't get stuck between violating organizational rules (which may make you vulnerable) and having to ask for any single deviation (which may slow you down).

Typical cases in the area of data are around data privacy or data security. Of course, you will use any granted freedom with the necessary caution.

# Learn from the crisis

## When does a crisis end?

Most crises do not come with a formal end date. You can usually expect the situation to improve over time. However, you should always consciously switch from crisis mode to normal mode. It is, therefore, useful to agree on end criteria during the crisis.

Returning to "normal mode" may not mean returning to the way the organization operated before the crisis. The crisis may have changed the organization, its industry (often triggered by technological progress), or even the entire world, as observed with both the financial crisis of 2008 and the COVID-19 crisis of 2020.

## The crisis as a catalyst

Crises often see innovations introduced which otherwise would have taken years to get the buy-in from all necessary stakeholders. As a lot of innovations are about data – notably in the area of digitalization – such exceptional situations are great opportunities.

The aftermath of a crisis is the time where organizations decide whether to keep such innovations or return to the old way of working.

At the same time, certain activities you had to stop during the crisis may have turned out not to be missed. I recall a myriad of reports we once had to stop just to find out that the original requestors had left the company years ago, and nobody has been reading them since.

You may wish to use the opportunity to abandon all activities that turned out to be dispensable – your team can direct their energy to activities that generate higher value.

## Lessons learned

During a crisis, you learn about dealing with a crisis, but you also learn a lot beyond the crisis. You learn about your organization, your industry, and your customers. It is, therefore, a good habit to not only document your "crisis lessons" but any kind of behavioral pattern you have discovered during that phase.

You may even find out that you require a different Data Office organization to adequately handle the postcrisis situation. If findings, together with first conclusions and concepts, become part of your "lessons learned" documentation, it may become easier to justify the changes – which is essential if you intend to enhance your portfolio of activities or responsibilities.

## Celebrate

As with every completed project, you should celebrate having overcome a crisis as a team. This is often the first opportunity to look back together. The tales being shared during the celebration ("Do you remember what we did when...?") are something people will usually keep in mind for good. It becomes part of the organization's legacy and creates a feeling of belonging. Loyalty is built on such experiences.

The ability to celebrate is another reason to consciously declare a crisis overcome while stressing that a lot of tidying-up activities remain. Without such a definite date, organizations are at risk of slipping back into the daily business without recognizing outstanding effort and achievements during the crisis.

During crises, many employees go beyond their usual dedication. You may want to keep this momentum. To do so, you should not reward people for what has been achieved – this would allow them to lean back and enjoy. Instead, praise their effort, their courage, and their can-do attitude. This approach generally motivates people to extend that behavior beyond any completed task or overcome crisis.

# Data in Mergers and Acquisitions

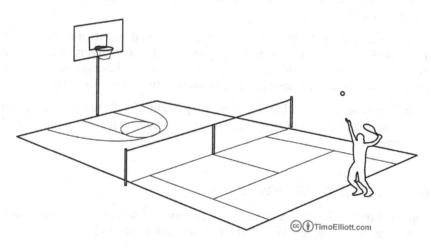

*Before using game-changing technology,*
*you have to know what game you're playing!*

**Figure 22-1.** Same size and same genre may not make it easier

© Martin Treder 2020
M. Treder, *The Chief Data Officer Management Handbook,*
https://doi.org/10.1007/978-1-4842-6115-6_22

# What is going wrong today?

Many mergers of different organizations – be it between equals or as an acquisition – fail.

Since mergers and acquisitions have become common in our free-market economy, the perception has developed that most failures are caused by a neglect of cultural and human aspects.

This perception has resulted in an increased focus by organizations around the world on understanding and addressing these cultural aspects of merger and acquisitions. This focus helps avoid many errors from the past.

Unfortunately, at the same time, this shift has caused organizations to underestimate the aspects of operative compatibility.

As a result, most of today's failures in integration cannot be attributed to cultural issues – as much as some consultants want to make us think this way.

Instead, more and more mergers fail for mundane **compatibility** reasons. And whether it is for incompatible sales areas or differently structured product portfolios, most nontechnical compatibility issues are around different handling of **data**.

As a condequence, data cannot be fixed as an afterthought of a merger or acquisition. Instead, all aspects of integrating two or more organizations need to be addressed from a business perspective, a technical perspective (where applicable), and a data perspective at the same time in a joint effort.

That is why, next to a plan for technical integration of different hardware and software worlds, successful integration requires a systematic, cross-functional data integration approach.

# Integration planning

A typical integration plan is based on a number of subject areas for which workgroups are to be set up that start their work in parallel.

Typical challenges are interdependencies between the workgroups, a rewarding scheme focused on short-term success, and negligence of data aspects.

An organization that finds itself in a merger situation would ideally start with the same logical sequence of steps as for any business change process, with a strong focus on data from the very beginning:

(i) **Market positioning**

What do we want the resulting organization to stand for? What would customers ideally want to buy from our organization? And why would they prefer us over our competitors?

(ii) **Target service offering**

What are our future products and services? How do we intend to structure and identify them? What is the intended level of complexity (e.g., a set of products defined individually by market or localized features)? To which extent can we structure the service offering, so that subsequent steps can be automated (e.g. production or invoicing)? Where, and to which extent, does our target service portfolio deviate from any of the previous portfolios?

(iii) **What are our target processes?**

Our service offering needs to be supported by processes. It must be determined to which extent these processes can be predefined and in which cases the individuality of a rendered service forces us to work with policies instead.

We should devise different alternatives, compare them, and estimate the impact of each of these options. We should be as precise as possible with regard to the determination of cost, time, and availability of resources, which are critical aspects for a selection of the target process in each case.

(iv) **What are the data requirements?**

How do we structure all aspects of data? Where do we have data model incompatibilities? Where can we match the two (or more) data models, where can we find a workaround, and where is it impossible to merge two worlds without loss of information, both in data repositories and daily operation?

The same steps as under (iii) apply! Different target data models need to be developed, and the same criteria need to be applied to select one target data model. This aspect is critical.

No software development must be allowed without a target data model to work toward. Interim models are possible and often even necessary – but the target model must always be known, and it must be given priority.

### (v) Technical implementation

Based on the results of (iii) and (iv), we need to develop a road map, taking the costs of change into consideration again. Quick wins are okay even where the pure cost calculation would not suggest them.

This is because of psychological aspects: People need to see results soon. They will not stay loyal if they are kept waiting for someone to switch on the final solution at the very end of the process. In addition, everybody can learn from working with interim solutions, and these learnings may help improve the target solution design.

Furthermore, work will be required on throwaway stuff, notably to obtain temporary compatibility between areas that proceed with different speed.[1]

Dependencies need to be mapped out between the schedules of all areas to determine the critical path. Hardly anything is as annoying as nine completed workstreams that are forced to wait for the tenth one.

It goes without saying that this approach is not meant to be considered a strict sequence. It must be seen as an iterative process. Hardly anything is as challenging to plan in advance as a merger, where people from either side face surprises every day.

## The data approach

Here are a few recommendations when planning the handling of data as part of a merger or an integration.

---

[1] This requires a big, overall business case so that such interim work can be justified as part of the long-term plan.

# Who should manage data integration?

Challenges should not be addressed by directly impacted business functions in isolation, independently of each other. Such an approach easily results in solutions that solve one team's problems while having adverse effects on other teams.

Each merger or acquisition should start with an assessment of data issues in the merging process. This activity will result in a long log of interoperability and reporting challenges, usually based on data mapping issues.

To address the subject of "Data in M&A" adequately, Data leadership should be clarified early. If a merger results in two co-existing legacy Data Offices, the decision for the single CDO should be taken early, while all members of both teams will be required for the work ahead. Until a final leadership decision is taken, try to agree with your peer on a "beauty contest under ceasefire conditions."

If you gain the lead, you should immediately set up a dedicated (temporary) entity responsible for the data aspects of the entire M&A program. If there is a dedicated budget for the merger or integration (there usually is), you should claim a fair part, so that you can address the data aspects without severely impacting the Data Office's daily work.

# Understand the motives

Whether you face integration or acquisition, it is vital to understand the drivers. This helps decide whether to integrate quickly or thoroughly (or to find a balanced approach in between the two extremes).

A long-planned strategic merger will more easily allow for a "get it right the first time" approach, whereas a survival merger of two players, both of which are too small to survive, will probably focus on "making it work somehow" in the first place.

Before you try to influence the decision with data-related consideration, it is indispensable to fully understand the underlying priorities of the Management Board, maybe even of a potential Supervisory Board.

But you may generally expect different motives behind an organization's decision to harmonize, whether triggered by a merger or by strategic or operational considerations.

Some organizations intend to integrate previously independent country organizations or entities so that they can be managed together, and their performance can be compared more easily.

Sometimes, organizations want to achieve a harmonized product offering across all entities, or a single brand should stand for the same value proposition.

Other organizations decide to harmonize so that they can outsource or automate standardized processes. (Experience shows that organizations should standardize **before** they outsource. If you outsource a variety of different procedures for the same activities, managing the supplier will become cumbersome, and the costs might go through the roof.[2])

In any of these cases, future business cases for data integration should usually be based on the main purpose of the underlying driver.

## Focus on interoperability

As part of any merger or acquisition, data from different parties must be brought together – be it for operational purposes, for a joint order-to-cash process, for customer visibility, or for financial reporting purposes.

That is why the primary integration target is *interoperability*, that is, the need for processes and/or systems of one side to be able to capture, interpret, and process data from the other side. An important aspect of interoperability is a harmonized identification logic for everything companies deal with, from raw materials to engines, products and employees.

This interoperability target must represent the "new world," not any of the two "old worlds." This means that, in most cases, the acquiring organization will need to change as well (unless it is a proportionally small acquisition).

You will also require mapping for joint reports. Typical business demand statements during an integration look like this:

- "There are some changes required in how Finance will report P&L of the acquired organization."

- "Who can tell me what data from one entity is already available in the data warehouse of the other entity?"

While a fully integrated organization will have these questions answered as a result of the integration, you will require additional effort to establish a proper reporting environment before and during the integration process.

Even where good progress is made in creating joint data repositories and in physically transmitting data between different entities, the **structure and meaning of data elements often differ tremendously!**

---

[2]A Deloitte survey has shown the #1 barrier to RPA: Process fragmentation – the way processes are managed in a wide range of methods – is seen by 36 percent of survey respondents as the main barrier to the adoption of intelligent automation. IT readiness is considered the main barrier by 17 percent of organizations (Deloitte, 2019).

And remember: Data challenges multiply if you pile up legacy issues on either side.

---

**Note** None of these challenges is specific to mergers and acquisitions. Harmonization of previously independent country organizations or entities may follow precisely the same approach. Technically, it doesn't matter too much whether the different entities have just become part of the same organization or whether they have already coexisted within the same organization for years.

---

## Create a high-level plan

No matter what is behind your organization's desire to integrate, experience shows that you can apply the following general sequence of steps:

- Determine best practices by topic in the different entities.
- Solve individual problems in a "reusable" way. Create general processes, guidelines, and so on.
- Determine best-suited processes and applications (which may come from a legal entity, a subsidiary, or a central function).
- Modify these processes and IT applications to be able to work together.
- Select one or two first entities to pilot the solution.
- Apply learnings from the pilot when rolling out globally.

## Determine the "best-of-breed" solutions

Each of the entities may be best at certain activities, or it may use the better models. You can use this situation to assemble a target setup that is better than each of the entities:

- Make a stocktaking of solutions and processes of all entities that are to be integrated.
- Measure them by the same standards to obtain visibility on which of the entities is best at which process.

- Determine the best existing solutions. Check them for compatibility and interoperability.

- Select the best valid combination for the target setup. You may have to compromise as different entities may be best at different processes, and those processes may not be fully interoperable.

## Don't innovate (too much) in parallel

Sometimes you will face the challenge that people will not agree to put innovation on hold for the entire integration period. Please resist. Insist on the principle of "harmonization before optimization." Harmonizing and optimizing at the same time is a recipe for failure.

Furthermore, a harmonization process that does not try to innovate at the same time will be faster, and keeping that process short is crucial.

Moreover, innovation and optimization will become far easier through successful integration, as you can focus on one single landscape of processes and data structures.

Finally, you will implicitly innovate wherever your integration follows the best-of-breed approach, as you are replacing solutions by superior ones.

## Data mapping

For successful integration of two or more different entities, you require a consistent view of all parts of the organization. To get there, you will need to map data fields and data elements.

You should use this opportunity and, wherever possible, work toward a normalized model which leads toward an agreed business target state.

Where such a state has not yet been defined, you should give it the highest priority – it is required for other integration activities as well.

Where such a definition is (technically or legally) impossible, your work should be based on agreed interim steps, without which no targeted merger activity would be possible at all.

A data mapping process is not a technical activity! It requires conscious business decisions. That is why it should never be done by IT alone.

The reason is that mapping is often ambiguous. Where you cannot find a perfect 1:1 mapping, you need to find a mapping that comes close. But which of the possible options should be taken? This decision should be based on business considerations: You need to take into account the immediate impact on your organization's business but also compliance with the final target.

Once all business decisions are taken, a close collaboration between the Data Office and the IT Architecture team is required to document the results in a format that allows for the unambiguous implementation of the underlying processes in IT applications.

Typical artifacts are data flow diagrams, entity-relationship models (for target and transitory steps), and a canonical data model.

# Organization

Your "Data in Integration" organization will need a single point of contact for all mapping needs. This person will typically not execute the mapping but coordinate and orchestrate all mapping activities. Strict coordination is vital to avoid duplication of work as well as gaps and inconsistencies in mapping between different functional initiatives.

All mapping initiatives coordinated by this entity will need to be subject to a **completeness check** from a data perspective, as your organization would do with any other integration activities. Here is a typical checklist:

- The business problem is stated.
- The Data Champion is determined and ready to take business ownership.
- Cross-functionality is addressed.
- Deliverables are clear (consider both business deliverables and data deliverables!).
- Underlying business decisions are taken.
- Dependencies are clear.
- Timeline is defined.
- Work is organized.
- An agile change process is in place.
- The "Data Principles" (see Chapter 6) are followed.

Once you have addressed all of these points, the "data mapping" entity should limit its activities to monitoring, stocktaking, and documenting.

## Concrete mapping cases

There is no "one size fits all"! Typical mapping types are

(i) **1:1 mapping**

This easiest of all mapping cases applies where single elements or structural aspects of data are called differently but have identical meanings.

The resulting task is a traditional **glossary** topic: You need to define leading expressions and alias names/synonyms. It still requires collaboration between various functions to ensure you make the best possible choice.

(ii) **1:n mapping**

A typical example of a 1:n mapping is a list of attributes represented by codes where one party has multiple codes for as many different variants, while the other one subsumes all of them under one code, as there is no operational difference.

Adequate mapping requires operational understanding. You need to work closely with all impacted business functions (e.g., Production and Product Management) to find the best mapping.

In some cases, you might even have to change operational processes to properly enable processing across entity borders, if you cannot avoid a loss of operationally necessary information by other means.

(iii) **n:m mapping**

Imagine the following situation: In two lists of attributes that need to be mapped, granularity is sometimes finer on one side, sometimes on the other side.

Compared to the previous case, you face the additional challenge that you cannot just decide for the side with the finer granularity and start from there. Instead, loss of visibility needs to be accepted or addressed.

A solution or mitigation of the situation would usually be achieved through workarounds during the integration process. The biggest challenge is typically an organization's limited willingness to invest in the modification of tools and processes that are scheduled to be retired soon after.

A typical solution is the rededication of existing but unused codes. This comes with a high risk of ambiguity or misunderstandings. That is why such cases need to be carefully documented, communicated, and, most importantly, applied consistently across all business functions.

### (iv) No code counterpart

In this case, one party uses codes or data elements that the other one does not know.

If such codes are required as part of the target setup, the organization could use this opportunity to introduce them already to the side that does not use them today.

If, however, it is supposed to disappear after integration, alignment with the product integration activities is key: Which services/capabilities and visibility will no longer be offered as part of which product/service, and as of when? These decisions will help understand whether a workaround or a change to existing solutions can be justified.

### (v) One-dimensional structural difference

Here, the same dimension is used but structured differently.

A typical example is the reporting calendar where the base period may differ. This can be mapped relatively easily where common denominators exist: If one side follows the calendar year for reporting, while the other side reports by fiscal year, you could use "monthly reporting" as a common basis.

The situation becomes more challenging when you look at weekly reporting (typically the operational view) vs. monthly reporting (a typical fiscal view). Possible ways of still mapping these two base periods require compromises or organizational changes.

### (vi) **Complex structural difference**

Some cases base on entirely different logical approaches so that data structures cannot be mapped easily.

A good example taken from the transport business is how "shipment" (or "consignment") is defined. This goes far beyond possibly using different terms. The meaning can differ in multiple dimensions: One piece vs. multipiece, life cycle, visibility, "billed together" versus "transported together", customs handling, and so on. There is no standard recipe for a proper mapping, but it is evident that cross-functional collaboration is required.

### (vii) **Speaking codes**

These are codes where one field contains information in a substring. Think, for example, of customer identifiers where the first two characters are the country code of the customer, or remember the codes for general ledger accounts where the first digit classifies the type.

A target model should never foresee "speaking codes." It should work with well-defined attributes instead. Where you face such a challenge today, you'd decompose the code and add attributes to the target canonical model (which you need to define anyway). Besides, the code structure needs to be specified as part of your Metadata. The plan to overcome speaking codes requires a long-term view, plus an exhaustive stocktaking of all applications and processes that are built on that assumption.

# Data for Innovation

"Innovate? No—we already tried that once. It didn't work out"

**Figure 23-1.** Not everybody embraces innovation

© Martin Treder 2020
M. Treder, *The Chief Data Officer Management Handbook,*
https://doi.org/10.1007/978-1-4842-6115-6_23

# How can data drive innovation?

## Demystifying innovation

As a first step, how about demystifying "innovation" a little?

(i) **Innovation is always disruptive?**

Innovation is not necessarily about "the next big thing." You don't need to design a perpetuum mobile or a process that turns iron into gold to be innovative. Gradually improving something can be very innovative as well, if it is based on a good idea.

People may refuse to call great ideas with small impact "innovation." Don't let this stop you innovate at all levels!

Innovation is an attitude. You should try to innovate every day.

(ii) **Big corporations cannot innovate?**

Innovation is not a matter of size.

The innovation capacity of employees is independent of the size of the organization they are working for.

Big organizations may, at times, be too bureaucratic to implement ideas quickly – but their maturity often avoids premature implementation. This helps concentrate on promising ideas – without being less innovative.

After all, innovation is primarily a matter of individual mindset, not of the size of the organization.

(iii) **Innovation is for creative specialists only?**

No, it is not a topic for professional "innovators" or innovation departments only. Even more importantly, innovation is not an unplannable result of arbitrary brain activity of creative people.

Creativity is indeed crucial for innovation, but this is as much a matter of techniques and organization as it is of natural talent.

Please expect innovation to be 10 percent great ideas and 90 percent hard work.

Figure 23-2 indicates how much of the innovation work needs to go into creating an innovation-friendly environment and into turning ideas into commercially effective innovation.

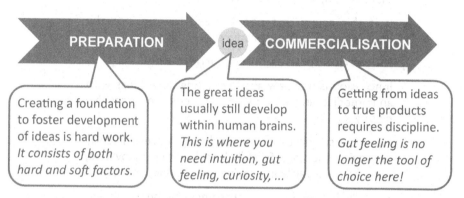

**PREPARATION**   idea   **COMMERCIALISATION**

Creating a foundation to foster development of ideas is hard work. *It consists of both hard and soft factors.*

The great ideas usually still develop within human brains. *This is where you need intuition, gut feeling, curiosity, ...*

Getting from ideas to true products requires discipline. *Gut feeling is no longer the tool of choice here!*

**Figure 23-2.** Innovation is more than a great idea

(iv) **Innovators must know their products?**

Many organizations innovate strongly around their products, with technically impressive results. But sophistication in, say, engineering or process design is not a guarantee for successful innovation.

Instead, organizations need to anticipate the thoughts, desires, and preferences of **customers and consumers**. This includes a good understanding of what makes them change their mind. After all, you want to understand *and* influence your customers.

But how about the need to understand your own business? Well, being too familiar with your current offering may even be counterproductive. You easily get stuck in a rut.

Steven Sasson, inventor of the digital camera at Kodak, is reported to have said: "Innovation best comes from people who really know nothing about the topic."

And Kodak itself has proven him right. Imagine how creative it was to invent a camera technology different from that of all existing cameras. And we all know how badly Kodak failed in turning this creativity into innovation. They thought they know what a camera has to look like: *You need a film behind the lens.* Really?

A key aspect of innovation is the ability to question **everything** you have in place today. This is not easy as it is emotionally challenging to scrutinize things you have assumed to be indispensable for years (or those that you may even have developed).

It is this ability that people need to gain, not the knowledge of existing products.

### (v) Innovation focuses on customers only?

Innovation is not only about creating new things with an immediate impact on what your customers receive from you.

Here is a simple example: Producing more efficiently or environmentally friendly can be very innovative, without necessarily improving the product as such.

Actually, impressing your customers by being innovative without producing tangible value for them may be a desired (side) effect.

But innovation can explicitly focus on the other relevant stakeholder groups, as well: Owners, suppliers, and employees.

- Any increase in "output value per input value" is usually appreciated by **shareholders**, leading to a rise in an organization's market valuation. Investors generally define "innovative" in a broader sense than mere customer satisfaction.

- **Suppliers** are not only a good target for innovation so that you can acquire your supplies and services more efficiently. They are also a great potential partner for collaborative innovation along the value chain.

- Innovation can make your organization more attractive to current and future **employees**, even without touching the quality of products or the effectiveness of their production. Better working conditions, clearer career paths, or better opportunities for personal development can all be achieved through innovation, unlike a mere rise in wages.

## (vi) **Innovation requires even more data?**

Today, innovators do not struggle with gathering or generating data. Most organizations are only using a small fraction of the huge amount of data they already have.

It is, therefore, advisable to first focus on exploring data that is already in your hands while still being unused. Secondly, a lot of data may already get captured but not used beyond supporting immediate operational activities. And there is a third group of data that could be collected at no or little cost.

So much innovation power lies in these sources of data alone that most organizations wouldn't even have the bandwidth to deal with additional, possibly acquired data.

There may be exceptions in concrete cases such as the acquisition of data you know you will need, but as a rule of thumb, organizations should concentrate on using available[1] data before piling up even more bits and bytes.

And, as you will remember, personal data should not be gathered on suspicion of future applicability anyway to avoid conflicts with privacy regulations such as GDPR.

# What is "data-driven innovation"?

If you think about data and innovation, what comes to mind first? Innovative methods of data handling, for example, of new applications of unsupervised learning or of multidomain Masterdata management in the Cloud?

All of this is both fascinating and relevant, but all too often it pushes another aspect into the background: Data Management as a **facilitator of business innovation**.

In other words, data-driven organizations would not change the people and processes for innovation in the first place. Instead, they would equip them with data and tools to ease their job.

---

[1]In this context, "available" refers to data that you have in your own databases as well as external data that you have free access to, i.e., free data or data you are already paying for.

And, indeed, tremendous opportunities lie in exploring conventional yet innovative ideas that are either based on data or later validated through data.

This is "data-driven innovation." It has been recognized as a dedicated discipline, and people have started to use the acronym **DDI** for it. The OECD published a definition as early as 2013.[2]

It is essential to consider that DDI is not competing with "conventional" innovation – it has a supporting role. The same R&D labs will keep working on innovative solutions, yet more and more powered by data.

## Using data to innovate

Data is not only a topic for operational support or forecasts. It can fuel innovation in multiple ways.

Data can, in particular, reveal correlations and causalities, within or outside an organization. It can also answer previously unasked questions by determining a previously unknown demand or a new product opportunity.

But data also offers alternative ways of tailoring a product or service, as it can help gauge customers' willingness to pay for new or different products.

After all, most organizations' ability to innovate is not limited by a lack of ideas – these are piling up wherever employees put themselves in the shoes of consumers. Instead, commercialization is often prevented by a considerable degree of uncertainty about their chances of commercial success, in other words, by risk-aversion.

But data is, of course, not limited to the validation of existing ideas. Even the development of new products or services can be supported by data, for example, through a simulation of millions of combinations of dozens of attributes or ingredients – doing so manually would take humans ages.

Furthermore, data can allow for new ways of advertising a product – by addressing people individually and demonstrating how an innovative product meets individually different needs.

But data can even support profane innovation such as the determination and validation of new production processes. Such processes allow for better products at the same price or for the same products to be produced at a lower price.

---

[2]See "Exploring Data-Driven Innovation as a New Source of Growth: Mapping the Policy Issues Raised by 'Big Data'" (OECD, 2013).

# Supporting data-driven innovation

Innovation doesn't happen if you hope for it to happen, of course, and organizations that don't innovate will be overtaken by others who do. That is why it is not sufficient to simply **allow** for data-driven innovation. Instead, organizations need to actively **create** a data-driven culture – not through posters and slogans, but by encouraging and rewarding each individuum.

But how can an organization effectively support innovation driven by data?

## Determining roadblocks

Data-driven innovation gets stuck for multiple reasons and at different levels in the organization.

(i) **The executives.**

Is innovation positive? Not everybody would agree! Executives who have helped shape a successful organization, sometimes over decades, may become very proud of the status quo and of what the organization has achieved in the past.

A second roadblock on the executive floor is the frequent confusion of efficiency gains and cost savings with innovation:

**Cost savings** will be copied by your competitors without delay (or maybe *your* organization is copying the competition?) so that you can hardly win the market through cost reduction alone.

**Innovation,** in contrast, is something your competitors cannot replicate easily.

(ii) **The workforce.**

The workforce is ready for a data-driven organization because Digital Natives are data-savvy.

Really?

To judge appropriately, it helps to realize that private data handling differs from business data handling.

Think of your smartphone. It does everything by a tap of a finger. You don't need to align with anybody.

Is this how you should deal with data in business?

I guess you should not, and this is because of the following difference:

- In private life, **somebody else** is ensuring that data helps **you**.

- In business life, **you** are expected to take care that data helps **others**.

In this sense, we could say that the people who **develop** those smartphones are data-savvy, not the users. How many of those smartphone developers do you know in your organization?

(iii) **Middle management.**

Would any manager in an organization openly admit being against a data-driven organization? Probably not!

In fact, you would hardly ever see open opposition to the idea of becoming data driven. Instead, the following pattern can be observed in many traditional organizations:

Step 1 **The Board pushes innovation.**

Being a public supporter of data and digital is common courtesy among executives. If you read interviews with CEOs in magazines like *Forbes*, you know what I mean...

Step 2 **The workforce loves it.**

If technology makes your private life easier every day, chances are you would welcome this kind of progress in the workplace as well! As long as it is not perceived as competition (putting your own job at risk), data-driven progress is generally considered positive.

Step 3 **But middle management has concerns.**

Whom am I talking about? All managers who are not involved in the development of the organization's strategy but who have teams of significant size (e.g., a branch manager). They are expected to carry the strategy into the field, as promoters and multipliers. But, more often than not, this is where innovation loses its momentum.

Are those managers fools? Do they fail in understanding?

No, most middle managers are sufficiently intelligent, and they usually like innovative ideas as well! But they have good, very individual reasons not to promote change.

When Rainer Meier was Head of Corporate Communications at Deutsche Post in 2004, he had already observed this phenomenon. He used to call it the "Cloud of Middle Management" which he identified as the management layer where the knowledge sits and where change is perceived as a threat (Meier, 2004).

So, where do these concerns come from? And what to do about this situation?

A Management Board that wants to become change embracing might wish to establish a channel to give its middle management a voice. Subsequently, patterns need to be discovered and reliable answers to critical (and usually valid) questions need to be developed. Here are a few typical examples of concerns when dealing with innovation:

- What if I fail? Do I have one shot only? Can I lose more than I can win?

- What's in it for me? Does it fit into the story of my career?

- Will innovation divert energy from my team's day jobs (the success which I am measured against)?

- What if I don't understand? After all, innovation often comes with new technology that I wasn't able to familiarize with at the university 30 years ago.

Successful organizations deal with such questions openly as part of their leadership development. Does your organization have such a program in place?

# Organize Innovation

A well-conceived framework helps fuel data-driven innovation. I consider the following three aspects critical to the creation of the necessary solid basis:

### (i) Creating a suited organizational structure

Taking an innovative idea forward is easier in a protected innovation space. Otherwise, innovation gets buried under a plethora of daily tasks.

If you want to enable your organization to innovate with data, you might prefer a data-centric Research and Development approach. After all, the same data sources may fuel innovation in various different product or service areas.

However, not all innovative data work needs to be taken out of the daily business and put into an R&D environment. Even within normal departments, you may wish to provide organizational support so that people can take an innovative idea forward without being distracted by administrative tasks.

If your organization is big enough, it is a good investment to have a team of data experts trained in turning ideas into reality. Those people can support colleagues from different departments who have promising ideas.

Sometimes, even the creative task of formulating an innovative approach and the processing of that approach require different skill sets. So, if someone has a great idea, don't force that person to do all subsequent steps as well. You should have people that are both trained *and* willing to take over.

(ii) **Managing the *entire* innovation process**

Some people tend to lean back prematurely, typically after an interim step has delivered an impressive result.

Examples for such results are a broadly admired concept, a signed-off piece of software, and a successful pilot.

You may want to prevent an innovation process from stopping here.

This is where a diligently described "innovation process" can help. It covers the entire chain from ideas and from research up to productization to ensure full attention is paid until the very last step.

As soon as an opportunity emerges, a good innovation process ensures that this opportunity is recognized, formalized, and provided with priority and resources – including the "ugly work," for example, the often tedious calculation of a business case.

Note that formalization does not necessarily kill creativity. Instead, a well-designed process prevents creative people from getting distracted while, at the same time, it ensures all necessary noncreative activities get done.

Furthermore, all relevant business functions need to get involved at an early stage. If a creative idea stays in a protected environment for too long, many critical, nontechnical side effects disappear from view.

---

## EXAMPLE I

Think of an organization that considers using drones for goods delivery purposes.

This creative idea alone is not an innovation!

It is relatively easy to pilot this concept, and the controlled environment of such a test does almost guarantee success.

Through such an initiative, an organization can shape its reputation as a creative driver of innovation, and the positive press articles are a given.

But then the hard work starts:

a) Customer focus: The organization needs to determine the cases in which drone delivery adds value to the customer. Why should a customer pay (more) for such a service? How many customers would go for such a product?

b) Commercial focus: The organization will have to calculate the added value and validate it through commercially oriented pilots. Does it earn us (more) money?

c) Operational focus: It will be indispensable to determine all possible deviations from the happy flow, such as vandalism or potential collisions between drones. Solutions need to be developed in engineering, process design, and software. The result needs to be operationalized, so that it works in daily routine and that it scales.

d) Legal focus: Compliance is critical in the increasingly regulated drone business – while new, as of yet ungoverned aspects need to be anticipated. Safety and security aspects need to be taken care of as well.

All in all, you can expect the effort up to the first successful pilot to cover less than ten percent of the entire work.

It is for good reasons that none of the prominent logistics corporations has established a big-scale drone business within the first two years after we saw the first photos of their branded delivery drones in the media.

---

(iii) **Making your data available**

You should organize your data so that all potential innovators can see what data is available to them.

And please give access to all of them. This access should include support from the Data Office. It should also come with all relevant information about the data, that is, the degree of reliability of its source, its age, its structure, and its metadata, but also its quality, for example, its completeness, format compliance, and so on.

It is helpful to consider that there may be more potential innovators in your organization than the limited number of data experts and research engineers.

## Proper handling of business cases

Business cases are indispensable before anything is added to an organization's product offering. However, business cases should not be applied too early during the innovation process, as this approach may kill a promising idea before exploring all options to make it profitable.

Where information is missing for a comprehensive business case, a two-step approach should be foreseen: Starting with a low-cost preproject to determine the commercial feasibility of the idea, before launching an implementation project in cases where the preproject confirms the innovative potential.

## Adding data to your culture of innovation

Even in organizations with a highly developed culture of innovation, people often don't think of "data" as an inspiring factor in innovation. They think of brainstorming, market analysis, idea management, and so on.

This is where an explicit culture of "Let's ask the data!" helps a lot.

Of course, data alone will never innovate, just as a race car alone cannot win races. But even the best race driver will welcome a faster car. So let's encourage the best innovators to use data to increase their chances of success!

To get everybody on board (and not only the "natural-born innovators"), you would definitely think of publicly rewarding the right behavior:

- Instead of just hoping for creative ideas to come and to turn into innovation, people should be actively using the offered toolkit.

- Innovators should have the courage to stop projects as soon as profitability cannot be reached anymore. This requires a strong culture of rewarding people for such a step, instead of considering them failures.

- "Number of ideas with a good story per department" can be used as a public metric. The same applies to all further steps in innovation, including those that are less attractive.

- During the early stages, people should be allowed to play around, for example, to find even better variants of their original idea. That is why trial and error should be encouraged and rewarded. You could achieve that through the creation of "playgrounds" and by formally dedicating working time to creativity. (Please consider it part of "R&D," not lost working time.)

Good examples will help people understand the power of data. These examples do not necessarily have to come from your own organization or your personal experience. They just need to be real and credible!

# Commercializing data ideas

According to a NewVantage survey in 2019 (Brown, 2019), a mere 11 percent of Chief Data Officers have revenue responsibility.

Many organizations may be missing an opportunity here, given the fact that more and more organizations build their entire business model on data.

Even where an organization's main objective is not directly based on commercializing data, a CDO should not have to wait for business people to ask for help.

Instead, the data itself may come with opportunities – internally or even as an external product offering.

# The "hundred thousand customers" strategy

- Information about one customer is anecdotal.

- Knowing 100 customers allows for first insight.

- Having 100,000 regular customers is a treasure from a data perspective.

If you work for an organization with such a huge number of customers, you may wish to develop a strategy to exploit this asset, even beyond your core business: People or organizations buying from your organization.

No other data area offers such great opportunities from combining your own information and external information than customer data: Social media knows a lot about your customers that they would never share with you directly.

---

## DATA MANAGEMENT THEOREM #12

**A hundred thousand data records are worth more than a hundred thousand times the value of one data record.**

---

Here are a few ideas for your "hundred thousand customers" strategy:

- A hundred thousand customers could be enough to create a community: People who exchange thoughts, questions, and ideas around your core offering – under the visible brand and logo of your organization.

- Those customers could allow you to collect data that no other organization (at least none outside your organization's industry) can gather. That data may be of vital interest for organizations in other sectors.

- Without having to deal with individual customers (remember GDPR), a sample of 100,000 anonymized records allows you to determine statistically relevant preferences and behavioral patterns – which you may even be able to sell: Other organizations are often interested in correlations and causalities on the attribute level, without requiring personalized data.

- You can attract those customers to other service offerings of yours (thus generating additional revenue), where their relationship with you gives them a starting bonus.

- You can extend your business vertically to make the customer experience seamless and straightforward. (Not only organizations as big as Amazon can do that!)

- You can find out about upselling or cross-selling opportunities based on statistical evaluation of transactional data instead of guessing.

- You can ask what your lost customers have in common. This allows you to move from customer loss analysis to the prevention of customer churn, by taking action on the determined groups before they leave.

- You can understand your customers more systematically through Analytics. A vinery would not only know how many of their customers buy enough white wine bottles for a wine fridge to potentially be of use. They would also be able to determine the percentage of customers who would be open to such offerings. Finally, they would even find out which combinations of customer attributes indicate a higher chance of buying auxiliary devices. Data from 100 customers would not be sufficient to find out, as many of the numerous combinations of attributes would be represented by one or two customers only.

- You can use your well-introduced brand for totally different products or services, and customers start their experience with trust inherited by your good brand.

- A poll of 100 people is statistically questionable. However, it is difficult to ask more people at a reasonable cost. Asking 100,000 existing customers gives you a solid statistical base. They are not representative of the entire population but possibly of your target group!

- You might gather data that is valuable for other organizations to whom you could sell it, even as raw data.

It is a considerable strength to be able to test new ideas over and over again if you have the size to do so. Small organizations may be agile, but big organizations have a broad customer base *plus* the money to keep trying. You only know that something does *not* work if your own organization has tried it. Others may have failed because they have tried it differently – which makes it risky to draw immediate conclusions for your own organization.

Try **weird** ideas. Looking back, most true innovations started this way. Assume all **ordinary** ideas to have been explored before. You won't ruin your reputation by trying and failing any more. Entrepreneurial courage is appreciated by customers these days.

## Data innovation factory

Have you ever witnessed the gap between laboratory and "going live"?

I have observed this industrialization gap most frequently with "data products" – both with internal products and products to be offered to external customers. They seem to be prone to staying "academic."

A data innovation factory may be the most sustainable way of organizing data-driven innovation, as described earlier in this chapter. It would be responsible for all steps from the first ideas to the piloting phase, product definition up to the initial assessment of the resulting offering.

Such a central body would also guarantee that different ideas are measured by the same standards. It can help remove the bias from prioritization.

This is your chance to "walk the talk": Ideas that are candidates for data innovation are to be judged and rated through DATA.

# List of Theorems

| Number | Theorem | Chapter |
|---|---|---|
| I | **Data needs governance.** The organization of data-related responsibilities and activities demands a careful balancing between centralization and delegation.<br><br>• Any centralization requires good reasons.<br>• Any delegation must involve trust and support. | 2 |
| 2 | **Data is a business job.** Data Management is everybody's job – throughout the entire organization. It is particularly not an IT job, and it does not start with technology.<br><br>It is about bridging the gaps, based on a solid understanding of both business and IT. | 2 |
| 3 | **Data uses ALL facts, always.** It is not sufficient to base decisions on some facts. You need to consider all relevant facts – across the entire organization. | 2 |
| 4 | **No Data without a mandate and buy-in.** A CDO requires both a Board mandate and the buy-in of the employees.<br><br>The former must be there from the beginning.<br><br>The latter must be achieved by the CDO. | 2 |

(continued)

© Martin Treder 2020
M. Treder, *The Chief Data Officer Management Handbook*,
https://doi.org/10.1007/978-1-4842-6115-6

| Number | Theorem | Chapter |
|--------|---------|---------|
| 5 | **The entire Data Supply Chain counts.** | 2 |
| | Data Management needs to cover all steps of the data supply chain, from the creation or acquisition of data up to its usage and disposal. | |
| 6 | **Data is cross-functional.** | 2 |
| | Data Management needs to be cross-functional – because data is cross-functional. | |
| 7 | **Data means change.** | 2 |
| | Becoming data driven requires change across the entire organization. | |
| 8 | **Data is for everybody.** | 3 |
| | Data handling is not a topic for a small group of experts only. Data-driven organizations need to upskill their entire workforce, and they need to provide them with access to data. | |
| 9 | **Data succeeds with global standards, applied locally.** | 6 |
| | Data Management requires centralized governance. | |
| | Execution should be delegated by default. | |
| 10 | **Data is an asset.** | 16 |
| | Organisations should treat data as an asset, regardless of whether it is considered an asset from a legal or tax perspective. | |
| 11 | **Decisions require data.** | 18 |
| | Effective Data Management enables conscious, well-informed decisions on all management levels. | |
| 12 | **The value of Data grows exponentially.** | 23 |
| | A hundred thousand data records are worth more than a hundred thousand times the value of one data record. | |

# Bibliography

Allen, K. (2019). Radical Simplicity. London, UK: Ebury Press.

BA-Support. (2019, July 18). Business Analytics for Managers. Retrieved December 10, 2019, from www.ba-support.com: www.ba-support.com/doc/stat/Content/anova/anova.htm

Baxter, M. (2019, April 23). The future of the CDO: Chief Data Officers need to sit near the top. Retrieved December 17, 2019, from information-age.com: www.information-age.com/future-role-cdo-data-scientist-123481892/#

Bean, R. (2018, January 29). The Chief Data Officer Dilemma. Retrieved December 17, 2019, from forbes.com: www.forbes.com/sites/ciocentral/2018/01/29/the-chief-data-officer-dilemma/#678dedee3896

Beatie, K. (2018, January 18). The Importance of Data Management in Recruitment. Retrieved February 9, 2020, from hanrec.com: https://hanrec.com/2018/01/18/the-importance-of-data-management-in-recruitment/

Brown, S. (2019, August 15). Make room in the executive suite: Here comes CDO 2.0. (M. S. Management, Editor) Retrieved December 18, 2019, from mitsloan.mit.edu: https://mitsloan.mit.edu/ideas-made-to-matter/make-room-executive-suite-here-comes-cdo-2-0

Davenport, T. H. (2017, May). What's your data strategy? Retrieved from Harvard Business Review: https://hbr.org/2017/05/whats-your-data-strategy

Delesalle, P., & Van Wesemael, T. (2019). Deloitte Global CPO Survey.

© Martin Treder 2020
M. Treder, The Chief Data Officer Management Handbook,
https://doi.org/10.1007/978-1-4842-6115-6

Deloitte. (2019, December 05). How Companies Are Using Intelligent Automation to Be More Innovative. Retrieved from *Harvard Business Review*: https://hbr.org/sponsored/2019/12/how-companies-are-using-intelligent-automation-to-be-more-innovative

Dykes, B. (2016, August 24). Data-driven success rests on the shoulders of a strong executive sponsor. Retrieved December 10, 2019, from Forbes.com: www.forbes.com/sites/brentdykes/2016/08/24/data-driven-success-rests-on-the-shoulders-of-a-strong-executive-sponsor/#31c68cb52233

EU Commission. (2019, November 27). What does the General Data Protection Regulation (GDPR) govern? Retrieved from Website of the European Commission: https://ec.europa.eu/info/law/law-topic/data-protection/reform/what-does-general-data-protection-regulation-gdpr-govern_en

European Commission. (2020, February 19). White Paper on Artificial Intelligence – A European approach to excellence and trust. Retrieved from https://ec.europa.eu: https://ec.europa.eu/info/sites/info/files/commission-white-paper-artificial-intelligence-feb2020_en.pdf

Gieselmann, H. (2020, February 1). IT-Sicherheit: Von Clowns und Affen. c't(4/2020), p. 3. Retrieved February 1, 2020

GoFair. (2019). FAIRification Process. Retrieved December 10, 2019, from go-fair.org: www.go-fair.org/fair-principles/fairification-process/

Goyvaerts, J. (2019, November 22). Regular-Expressions.info. Retrieved December 25, 2019, from www.regular-expressions.info/: https://www.regular-expressions.info/

Hellard, B. (2019, September 30). AI and facial analysis used in job interviews for the "first time". Retrieved December 12, 2019, from ITPro.: www.itpro.co.uk/business-strategy/careers-training/34522/ai-and-facial-analysis-used-in-job-interviews-for-the-first

IFRS. (2017). IAS 38 Intangible Assets. Retrieved from IFRS: www.ifrs.org/issued-standards/list-of-standards/ias-38-intangible-assets/

InformationAge. (2006, December 22). Intelligence as a service. Retrieved November 14, 2019, from information-age.com: www.information-age.com/intelligence-as-a-service-276141

Labovitz, G., Chang, Y., & Rosansky, V. (1993, 11 01). *Making Quality Work: A Leadership Guide for the Results-Driven Manager*. Wiley.

Logan, V. (2019, March 07). Be the centre of gravity not control. Retrieved from Information-Age.com: www.information-age.com/gartners-chief-data-officer-survey-123480481

Luke. (1971). Holy Bible – English Standard Version. In Holy Bible – English Standard Version. Crossway.

Marr, B. (2019). Why every business needs a data and analytics strategy. Retrieved December 10, 2019, from bernardmarr.com: www.bernardmarr. com/default.asp?contentID=768

Meier, R. (2004, May 3). Communication in Time of Change. DHL Advanced Business Leadership Programme. Boston, MA, USA.

Möller, A. (2019, November 25). Heading AI lighthouse cases at Bayer. Retrieved December 3, 2019, from LinkedIn.com.

Moran, M., & Logan, V. (2018, June 25). Success Patterns of CDOs Driving Business. Retrieved from Gartner.com.

OECD. (2013, June 18). Exploring Data-Driven Innovation as a New Source of Growth: Mapping the Policy Issues Raised by "Big Data". OECD. Paris: OECD Publishing. doi: https://doi.org/10.1787/5k47zw3fcp43-en

Parker, S., & Walker, S. (2019, 10 28). Think Big, Start Small, Be Prepared – Master Data Management. Retrieved November 14, 2019, from gartner.com: www.gartner.com/doc/reprints?id=1-1XTK8FUK&ct=191127&st=sb

TOGAF. (2019, 12 01). The Open Group. (S. Nunn, Editor) Retrieved December 25, 2019, from www.opengroup.org/togaf

Treder, M. (2012, March 31). Basics of Label and Identifier. Retrieved from SlideShare: www.slideshare.net/martintreder16/2012-03-basics-of-label-and-identifier

Treder, M. (2012, November 30). License Plate – The ISO Standard For Transport Package Identifiers. Retrieved from SlideShare: https://de.slideshare.net/martintreder16/license-plate-the-iso-standard-for-transport-package-identifiers

Treder, M. (2019). Becoming a data-driven organisation. Heidelberg, Germany: Springer Vieweg.

UPU. (2019, December 1). About Postcodes. Retrieved December 24, 2019, from www.upu.int: www.upu.int/en/resources/postcodes/about-postcodes.html

Wikipedia – Boiling Frog. (2019, October 13). Retrieved November 30, 2019, from wikipedia.org: https://en.wikipedia.org/wiki/Boiling_frog

Wikiquote – Helmuth von Moltke the Elder. (2019, December 16). Retrieved from https://en.wikiquote.org/: https://en.wikiquote.org/wiki/Helmuth_von_Moltke_the_Elder

# I

# Index

© Martin Treder 2020
M. Treder, *The Chief Data Officer Management Handbook*,
https://doi.org/10.1007/978-1-4842-6115-6

Printed in the United States
By Bookmasters